Medical Imaging of Normal and Pathologic Anatomy

Medical Imaging of Normal and Pathologic Anatomy

Joel A. Vilensky, PhD
School of Medicine
Indiana University
Fort Wayne, Indiana

Edward C. Weber, DO
The Imaging Center
Fort Wayne, Indiana

Thomas E. Sarosi, MD
Fort Wayne Radiology
Fort Wayne, Indiana

Stephen W. Carmichael, PhD, DSc
Mayo Clinic
Rochester, Minnesota

SAUNDERS

ELSEVIER

SAUNDERS
ELSEVIER

1600 John F. Kennedy Blvd.
Ste 1800
Philadelphia, PA 19103-2899

MEDICAL IMAGING OF NORMAL AND PATHOLOGIC ANATOMY   ISBN: 978-1-4377-0634-5

Notice

Knowledge and best practice in this field are constantly changing. As new research and experience broaden our knowledge, changes in practice, treatment and drug therapy may become necessary or appropriate. Readers are advised to check the most current information provided (i) on procedures featured or (ii) by the manufacturer of each product to be administered, to verify the recommended dose or formula, the method and duration of administration, and contraindications. It is the responsibility of the practitioner, relying on his or her own experience and knowledge of the patient, to make diagnoses, to determine dosages and the best treatment for each individual patient, and to take all appropriate safety precautions. To the fullest extent of the law, neither the Publisher nor the Authors assume any liability for any injury and/or damage to persons or property arising out of or related to any use of the material contained in this book.

The Publisher

Library of Congress Cataloging-in-Publication Data
Vilensky, Joel A.
  Medical imaging of normal and pathologic anatomy / Joel A. Vilensky, Edward
C. Weber, Thomas E. Sarosi ; Stephen W. Carmichael, editor-in-chief. – 1st ed.
      p. ; cm.
  ISBN 978-1-4377-0634-5
1. Diagnostic imaging. 2. Anatomy, Pathological. 3. Human anatomy. I.
Weber, Edward C., II. Sarosi, Thomas E. III. Carmichael, Stephen W. IV. Title.
  [DNLM: 1. Diagnostic Imaging–methods–Atlases. 2. Anatomy–Atlases. 3.
Pathology–Atlases. WN 17 V699m 2011]
  RC78.7.D53V55 2011
  616.07'54–dc22
                                                                    2009045679

Acquisitions Editor: Madelene Hyde
Developmental Editor: Christine Abshire
Publishing Services Manager: Anitha Raj
Project Manager: Beula Christopher
Design Direction: Louis Forgione

Contents

Foreword vii
Acknowledgments ix
Introduction xi

Head and Neck

1. Hydrocephalus (MRI) 1
2. Cephalhematoma (CT) 2
3. Metastatic brain tumors (MRI) 3
4. Primary brain tumor (MRI) 4
5. Pituitary tumor (MRI) 5
6. Pineal gland cyst (MRI) 6
7. Papilledema—pseudotumor cerebri (MRI) 7
8. Vestibulocochlear nerve schwannoma (MRI) . 8
9. Acute epidural hematoma (CT) 9
10. Acute subdural hematoma (CT) 10
11. Chronic subdural hematoma (CT) 11
12. Meningioma (MRI) 12
13. Ischemic stroke (CT) 13
14. Internal carotid artery aneurysm (1)
    (angiogram) 14
15. Internal carotid artery aneurysm (2) (CT) ... 15
16. Carotid bifurcation plaque (CT) 16
17. Soft plaque, internal carotid artery (CT) 17
18. Maxillary and ethmoidal sinusitis (CT) 18
19. Asymmetry of the frontal sinuses (CT) 19
20. Blow-out fractures (CT) 20
21. Deviated nasal septum (CT) 21
22. Nasal bone fracture (CT) 22
23. Dislocation of the temporomandibular
    joint articular disc (MRI) 23
24. Degenerative joint disease,
    temporomandibular joint (TMJ) (CT) 24
25. Parotid gland tumor (CT) 25
26. Dilated submandibular duct, with
    calculus (CT) 26
27. Mandibular fracture (Panorex) 27
28. Basal skull fracture (CT) 28
29. Pharyngeal mass (CT) 29
30. Tongue (lingual) cancer (MRI) 30
31. Enlarged deep cervical lymph nodes (CT) ... 31
32. Thyroid nodule (US) 32
33. Thyroglossal duct cyst (CT) 33
34. Goiter (enlarged thyroid gland) (US) 34

Thorax

35. Pectus carinatum (CT) 35
36. Pectus excavatum (CT/radiograph) 36
37. Pneumothorax (radiograph) 37
38. Pneumonia (radiograph) 38
39. Pulmonary embolism (CT) 39
40. Breast cancer (mammogram) 40
41. Breast cyst, breast cancer (US) 41
42. Mediastinal tumor (CT) 42

43. Mediastinal lymphoma (CT) 43
44. Aneurysm of the ascending aorta
    (radiograph/CT) 44
45. Situs inversus (radiograph) 45
46. Right aortic arch (radiograph) 46
47. Coarctation of the aorta (CT) 47
48. Aberrant right subclavian artery (CT) 48
49. Coronary artery disease (CT) 49
50. Aberrant right coronary artery (CT) 50
51. Coronary angioplasty (CT) 51
52. Aortic valve stenosis (CT) 52
53. Atrial septal defect (ostium secundum)
    (MRI) 53
54. Hypertrophic cardiomyopathy (MRI) 54
55. Internal mammary (thoracic) artery
    coronary bypass (CT) 55
56. Pleural effusion (1) (radiograph) 56
57. Pleural effusion (2) (CT) 57
58. Emphysema (CT) 58
59. Lung cancer (radiograph) 59
60. Lung cancer, advanced (radiograph) 60
61. Lung cancer, right upper lobe (CT) 61
62. Large sliding hiatal hernia (radiograph) 62
63. Small sliding hiatal hernia (radiograph) 63
64. Esophageal varices (CT) 64
65. Diaphragmatic hernia (1) (radiograph) 65
66. Diaphragmatic hernia (2) (CT) 66

Abdomen

67. Metastases (CT) 67
68. Umbilical hernia (CT) 68
69. Inguinal hernia (CT) 69
70. Caput medusae (CT) 70
71. Ascites (CT) 71
72. Abdominal adenopathy (MRI) 72
73. Abdominal aortic aneurysm (CT) 73
74. Psoas abscess (CT) 74
75. Carcinoma of gastroesophageal
    junction (CT) 75
76. Duodenal ulcer (radiograph) 76
77. Ileal (Meckel's) diverticulum (fluoroscopy) . 77
78. Hepatic cirrhosis (CT) 78
79. Splenomegaly (CT) 79
80. Renal cyst (simple) (CT) 80
81. Renal cyst (complex) (MRI) 81
82. Urolithiasis, renal calculus (CT) 82
83. Renal carcinoma (US/CT) 83
84. Adult polycystic kidney disease/
    transplant (MRI) 84
85. Adenocarcinoma of the pancreas (CT) 85
86. Malrotation of the small bowel (radiograph) 86
87. Obstructed common bile duct (US) 87
88. Gallstones (US) 88
89. Volvulus (CT) 89
90. Appendicitis (CT) 90

91. Inflammatory bowel disease, regional enteritis, Crohn's disease (CT) 91
92. Ulcerative colitis (CT) 92
93. Urolithiasis, ureteral calculi, and dilated renal collection system (CT) 93

Pelvis and Perineum

94. Benign prostatic hypertrophy (BPH) (CT) 94
95. Uterine fibroids (Leiomyomas) (MRI) 95
96. Bicornuate uterus (MRI) 96
97. Ovarian cyst (US) 97
98. Ovarian dermoid cyst (teratoma) (CT/radiograph) 98
99. Urinary bladder diverticulum (CT) 99
100. Urolithiasis, bladder calculus (CT) 100
101. Varicocele (US) 101
102. Epididymitis (US) 102
103. Epididymal cyst (US) 103
104. Hydrocele (US) 104
105. Testicular tumor (US) 105
106. Testicular torsion (US) 106

Back

107. Axis (C2) fracture (CT) 107
108. Cervical intervertebral disc herniation (MRI) 108
109. Degenerative joint disease, cervical facet joints (CT) 109
110. Vertebral body compression fracture (CT) 110
111. Fracture of the pars interarticularis (CT) 111
112. Spondylolisthesis (secondary to pars defect) (CT) 112
113. Degenerative spondylolisthesis (1) (MRI) 113
114. Degenerative spondylolisthesis (2) (MRI) 114
115. Infectious discitis/vertebral osteomyelitis (CT) 115
116. Variation in number of lumbar vertebrae (radiograph) 116
117. Sacroiliitis (CT) 117
118. Herniated lumbar disc with neural compression (MRI) 118
119. Lumbar spinal canal stenosis (MRI) 119
120. Complete transection of the spinal cord (MRI) 120

Upper Limb

121. Acromioclavicular joint separation (radiograph) 121
122. Anterior shoulder dislocation (AP view) (radiograph) 122
123. Anterior shoulder dislocation ("Y" view) (radiograph) 123
124. Fractured rim of glenoid fossa (CT reconstruction) 124

125. Rotator cuff (supraspinatus) tear (MRI) 125
126. Superior labrum, anterior to posterior (SLAP) tear (MRI) 126
127. Enlarged axillary nodes (CT) 127
128. Dislocated biceps brachii tendon (MRI) 128
129. Olecranon fracture (radiograph) 129
130. Fracture of the radial head (radiograph/CT) 130
131. Pronator teres muscle tear (MRI) 131
132. Scaphoid fracture (MRI) 132
133. Triangular fibrocartilage complex (TFCC; articular disc) tear (MRI) 133
134. Colles fracture (radiograph) 134
135. Smith fracture (radiograph) 135
136. Boxer's fracture (radiograph) 136

Lower Limb

137. Posterior hip dislocation with fracture of the acetabulum (CT) 137
138. Metastatic tumor of acetabulum (CT) 138
139. Fracture of the proximal femur (radiograph) 139
140. Degenerative joint disease, hip (radiograph) 140
141. Avascular necrosis (AVN) of the femoral head (MRI) 141
142. Iliopsoas bursitis (MRI) 142
143. Obstructed femoral artery (CT arteriogram) 143
144. Deep venous thrombosis (US) 144
145. Knee joint effusion (MRI) 145
146. Medial (tibial) collateral ligament (MCL) tear (MRI) 146
147. Medial meniscal tear (MRI) 147
148. Quadriceps tendon tear (MRI) 148
149. Patellar tendon tear (MRI) 149
150. Anterior cruciate ligament tear (MRI) 150
151. Popliteal (Baker) cyst (MRI) 151
152. Degenerative joint disease, knee (radiograph) 152
153. Tibial fracture (radiograph) 153
154. Pes anserinus bursitis (MRI) 154
155. Calcaneal tendon tear (MRI) 155
156. Calcaneal fracture (CT) 156
157. Ankle fracture (radiograph) 157
158. Fracture of the medial malleolus and distal fibula (radiograph) 158
159. Ankle sprain (MRI) 159
160. Degenerative cystic changes, sesamoid bone of hallux (CT) 160
161. Plantar fasciitis (MRI) 161

Other

162. Radionuclide bone scan (nuclear) 162

Index 163

I was already the proud father of three beautiful girls when my pregnant wife went in for an ultrasound. She hadn't felt any movement for a couple of days, and the doctor was concerned. She remembers lying there in the dark in a polka dotted johnnie staring nervously at the ceiling. At last the technician spoke to her, pointing to the screen.

"See that?" she said, moving a little arrow toward exhibit "A"–an amorphous pale blob attached to a larger amorphous pale blob in a snow storm.

"It's a penis."

Our son came into the world screaming and healthy a few months later blissfully unaware of the intrusion into his privacy.

Our second encounter with medical imaging, in this case a good old x-ray, was produced about two years later and also relates to our son. It documents in no uncertain terms the results of a brief flight between the second from the bottom step and the floor. While we didn't have a picture of an intact clavicle for comparison, the break in the elegant curvature, combined with the scream each time we lifted him, left little doubt as to the problem.

At six years old, that same boy experienced the thumping and banging pleasures of an MR imaging machine as we attempted to precisely locate a splinter that had virtually disappeared into the depths of his fleshy sole.

While it may be starting to look like medical imaging technology has been developed exclusively for one boy, I offer these examples as a reminder of just how pervasive this technology has become. The fantastic leaps, excluding my son's, that have been made over the past few years have added several powerful weapons to the diagnostic arsenal by giving those responsible for our care unprecedented and generally painless access to our bodies.

While the term "medical imaging" may be new, the making of medical images certainly isn't. The work of countless medical illustrators has long provided the foundation for our understanding of human anatomy. Working from the real thing, whether in the morgue or the operating room, medical illustrators have been our guides to the complexity of the human body. In their roles as visual communicators, they understand that simply recording everything in front of them doesn't necessarily translate into useful information. Never losing sight of the purpose of each image they make, they edit as they go, focusing on the essentials while minimizing or eliminating that which is unnecessary.

The best medical illustrations are often stunning works of art as well as utilitarian documents. But they all, whether spectacular or not, are the result of careful observation based on years of experience, well-honed technical skill, and a well-developed editorial eye. And because they are made by hand, they naturally take time to produce. And time of course is the one thing we no longer seem to have.

Radiographs, CT scans, MR images and ultrasound scans are comparatively quick to produce, but because they are machine made, these "pictures" make no attempt to go beyond simply presenting the information they capture. In fact, they are so foreign to our experience, at least at first, that they become the subject matter rather than the body they represent.

It might be presumed by those of us outside the medical profession that the mere existence of these kinds of images will tell the experts what they need to know. But we would never assume that medical illustrators just somehow know what to do when they sit down in front of a body even though, like most of us, they've grown up with recognizable images– drawings, paintings, photographs, and videos– often with familiar landmarks, a tree, a person, not to mention easily distinguishable solids and shadows.

This more recent medical imagery, however, does not benefit from the same familiarity. By their very nature, these are pictures of things we don't see while growing up, at least on a regular basis, and for that reason they remain ambiguous to the untrained eye. They have to be translated for us. In the same way the body is turned into something accessible and enlightening by the illustrator, so must the messages hidden in the grays and grains of these remarkable images be made understandable if they are to meet their great potential. Whether it is the surgeon who will be guided by this information or the radiologist who provides it, both must be able to reconstruct from two dimensions the three-dimensional pulsating reality waiting nervously in the office or on the table, and they must be able to do this while constantly facing that other dimension–time. And that's where the building of "experience" needs something of a jumpstart.

When we look at anything for the first time, we may not know what to make of it. But eventually, with repeated viewings, over time we gradually learn to see differences, to identify growth, transformation, disappearance and even danger. Learning to recognize change is both a source of endless curiosity and an essential tool for the survival of all species. But we aren't talking about all species here. We're concerned with just one patient at a time. Being able to scan any area of a patient's body is remarkable, but if we don't know what that area is supposed to look like in the first place, how would we know if it's changed?

Here's a simple idea. Why not gather together pairs of actual images of the same area of the body and put them side by side? One to show the anatomy in its normal state, the other to present that same anatomy in its changed or abnormal state. Then, without obliterating the images, provide just enough labels and leader lines to know exactly what you're looking at. And finally offer a brief and yet completely adequate description of the particular pathology. Repeat this idea over and over again, while moving all around the body, and you just might find yourself not only seeing the differences more quickly, but also recognizing their significance.

No need to take my word for it, however. Just turn the page.

David Macaulay

Acknowledgments

We want to begin by expressing our gratitude to Elsevier for accepting our book proposal for this atlas. We especially want to thank Ms. Anne Lenehan and Ms. Madelene Hyde for shepherding the proposal through the Elsevier book approval process and for encouragement during its development. Ms. Christine Abshire was our developmental editor at Elsevier and is thanked for her skill and assistance on this project.

The actual process of converting our initial PowerPoint images to the professional quality plates that appear here was done by Mr. Chris Oakes of Graphic World Inc. in St. Louis, who tolerated us for two days as we asked him to move numerous leaders and labels microns so that the plates were printed exactly the way we wanted them to be seen. We also want to thank Ms. Cindy Geiss, Ms. Kate Challans and Mr. Mark Lane of Graphic World for helping to make our visit there fun and very productive.

We are very grateful for the dedicated efforts of Ms. Beula Christopher, who did the final conversion of our electronic files to the printed page. And we thank Dr. Diana Patterson for helping us check the proofs of this book.

Approximately half of the images in this atlas were produced using the facilities at The Imaging Center in Fort Wayne, and we are very grateful to its founder, Dr. Robert Connor, for his support and encouragement. A similar number of images for this atlas were obtained using the facilities of Fort Wayne Radiology and Parkview Memorial Hospital in Fort Wayne. We are grateful to the hospital and the physicians of Fort Wayne Radiology for allowing us to use these facilities, and especially to Mr. Chaz Eartly for his technical expertise and to Mr. Gary Stuby, department administrator of diagnostic imaging at the hospital, for his support. We thank all the technologists in both facilities for their work in producing the especially distinct images we needed for this book.

Dr. Scott Mattson, Medical Director, Echocardiology Laboratory, Cardiovascular MRI Laboratory, Lutheran Hospital of Indiana, graciously provided us with some of the cardiac images, and we very much appreciate his efforts on our behalf.

We are grateful to Dr. Chandana Lall, Department of Radiology, Indiana University Medical Center, for the CT image of pancreatic cancer.

Ms. Lowene Stipp, administrative assistant at the Indiana University School of Medicine-Fort Wayne, assisted us with many of the administrative tasks associated with assembling this book, and we thank her for this work.

We are honored that Mr. David Macaulay agreed to write the Foreword and want to thank him for doing so.

We are grateful to our wives, Deborah Meyer-Vilensky, Ellen Weber, Nancy Sarosi and Susan Stoddard, for their love and patience during the development and implementation of this project.

Finally, we thank the medical students who taught us how to teach and how to create anatomy/radiology books that both interest and instruct them in the beauty and complexity of human anatomy.

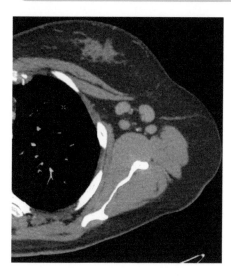

The axial CT image of the axilla *(above)* is from a patient with suspicious findings. A radiologist, upon viewing this image on the screen, would instantly recognize those findings and search for corroborating abnormalities on other images. The radiologist is alerted because he/she compares the above image with a stored mental image of a normal axilla. This knowledge of normal radiologic anatomy was acquired by the radiologist during residency and subsequent years of practice. Training and practice resulted in the skill of *pattern recognition* needed for that image to be meaningful.

With the advent of modern diagnostic imaging technologies (e.g., CT, MR, ultrasound), the role of radiologic imaging has increased tremendously in medical education so that now even first-year medical students in gross anatomy are expected to identify normal and pathologic anatomic structures on plain, MR and CT images. But these students don't have the benefit of years of examining patients with normal anatomy. This is the reason we created this book.

This second image to the right is a matched "normal" for the abnormal CT that is shown above. By "matched" we mean that the axial sections are at the same level in two patients with similar body habitus. Note that the second image does not show the four globular structures in the axilla that are distinctive in the first image. Those are abnormal axillary lymph nodes. Normal axillary nodes are typically much smaller and/or not as homogenous because they consist of tissues with different CT densities. The prominence and appearance of the abnormal nodes on the first image prompt the radiologist to look more carefully at the contralateral axilla, the mediastinum, the pulmonary hila, and in the lower neck for additional suspicious lymph nodes. The radiologist will look for other imaging findings

such as a lung mass, as well as review available clinical information that might explain the enlarged lymph nodes. Further studies or procedures, such as a biopsy, might be indicated to discern a specific diagnosis. An elaborate diagnostic process is thus initiated by the almost instant *pattern recognition* of an abnormal appearing axilla.

Medical students are at the beginning of the long process of developing the ability to quickly recognize the anatomic differences between health and when disease or trauma results in morphologic derangement. Existing anatomy illustrations do not typically aid the student's ability to rapidly perceive these morphologic changes. In this atlas we provide visual clarity of these differences by providing side-by-side images of the normal and pathologic condition in patients of similar body habitus, or highlight differences in the ipsilateral and contralateral sides. In this way we facilitate the acquisition of *pattern recognition* skills in students.

We include here images of pathologic conditions that have direct gross anatomic correlations and that can be visualized using routine procedures including radiographs, CT, MR, and ultrasound. These conditions typically are also discussed in two widely utilized texts in medical gross anatomy, *Gray's Anatomy for Students,* 2nd edition, and *Clinically Oriented Anatomy,* 6th edition, and we therefore provide specific page references to these texts.

Our labeling of structures in the images (most important structures are in **bold**) will enable the beginning student to comprehend the region. We do not present an explanation of imaging techniques (e.g., CT, MR) because such explanations are available in numerous texts and atlases, as is information on the standard orientations for radiographic images. However, for the images displayed we do indicate the

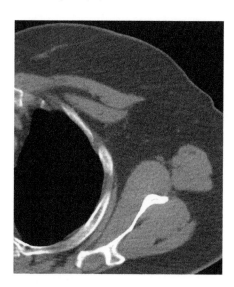

imaging modality (e.g., MR vs. CT) and where appropriate provide a brief explanation as to why this technique is typically the one used to evaluate the specific clinical condition. And we also often provide limited clinical information, although we recommend that the student refer to the pages listed for the anatomy texts or our previous book, *Netter's Concise Radiologic Anatomy*, for more comprehensive clinical information.

The student should be aware that images from ultrasound procedures are not as intuitive as are plain films and CT and MR images. Unlike these modalities, ultrasound images often show only limited non-orthogonal sections of anatomy. As illustrated in some of our ultrasound images, intensity of blood flow may also be represented on ultrasound images by arbitrary colors using the Doppler principle (for example, see page 144).

In some cases images for this atlas were obtained from an imaging modality that may not be the one most commonly used for that clinical condition because it displays the anatomy of a specific abnormality very well (see page 33). In addition, the images in this atlas were selected partly because only one anatomic derangement is apparent; in clinical situations patients often have multiple abnormalities.

The vast majority of the images in this atlas are from routine imaging procedures done at the Imaging Center (Fort Wayne, IN) and the radiologic facilities within the practice of Fort Wayne Radiology, at which two of us (ECW and TES) are practicing radiologists. A few images were obtained from other sources and the pertinent attributions are noted on the associated plates. We wish to emphasize that no patient underwent any additional (i.e., not clinically necessary) radiologic procedure in order to obtain the images used in this atlas.

During the preparation of this book, the authors had the opportunity to meet Mr. David Macaulay at the 2008 meeting of the American Association of Clinical Anatomists. He had just completed, *The Way We Work: Getting to Know the Amazing Human Body*. Similar to his previous books illustrating the building and operational principles of mechanical machines, common and uncommon, this work provides an outstanding introduction to human form and function. Mr. Macaulay's skills as an author and illustrator make the complicated seem (relatively) simple and we are honored that he agreed to write a Foreword for this book, which visually reveals how disease alters normal human anatomy.

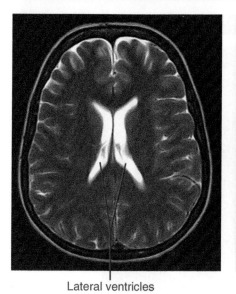
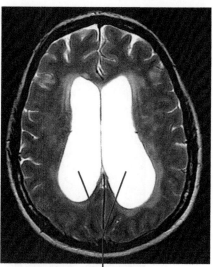

Lateral ventricles   **Dilated lateral ventricles**

Axial MR images of the normal brain *(left)* and one with hydrocephalus *(right)* at the level of the lateral ventricles. Hydrocephalus can result from excess cerebrospinal fluid (CSF) production, interference with the absorption of CSF (communicating hydrocephalus), or obstruction of CSF flow (obstructive or noncommunicating hydrocephalus) from the ventricles.

Gray's Anatomy for Students, 2e: Hydrocephalus (p. 834)
Clinically Oriented Anatomy, 6e: Hydrocephalus (p. 885)

Cephalhematoma

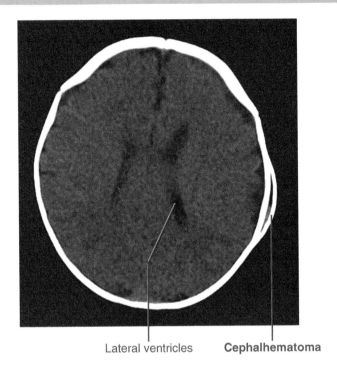

Lateral ventricles   **Cephalhematoma**

Axial CT image of the cranium of an infant. A cephalhematoma is an accumulation of blood between the skull and the periosteum of a newborn and often is a consequence of a forceps-assisted delivery. The hematoma usually calcifies, and then progresses to more mature ossification, as shown here. Parents are often quite concerned with the easily palpable "bump" on the skull, but can be reassured that this condition always resolves; the lesion is resorbed and disappears.

COA: Cephalhematoma (p. 861)

Metastatic tumors

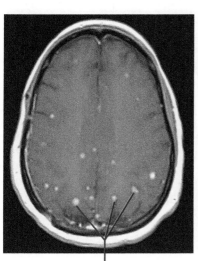

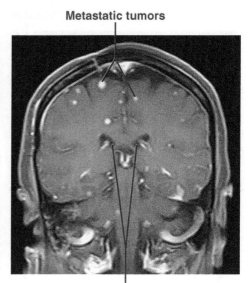

Metastatic tumors   Lateral ventricles

Contrast-enhanced axial *(left)* and coronal *(right)* MR images of the same patient. The bright round lesions within the brain parenchyma are metastases (five are labeled). These tumors are visible because of the leakage of contrast material from the blood into the interstitial spaces of the tumors. This is unlike normal brain parenchyma in which the blood-brain barrier limits the size and types of molecules that can leave the intravascular space.

GAS: Brain tumors (p. 835)
COA: Metastasis of tumor cells to dural venous sinuses (p. 876)

Primary Brain Tumor

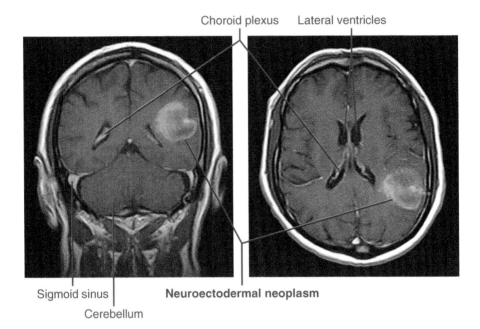

Choroid plexus   Lateral ventricles

Sigmoid sinus   **Neuroectodermal neoplasm**

Cerebellum

Contrast-enhanced coronal *(left)* and axial *(right)* MR images demonstrate a neuroectodermal (primary brain) tumor in the left parietal lobe. The patient presented with confusion and slurred speech.

GAS: Brain tumors (p. 835)

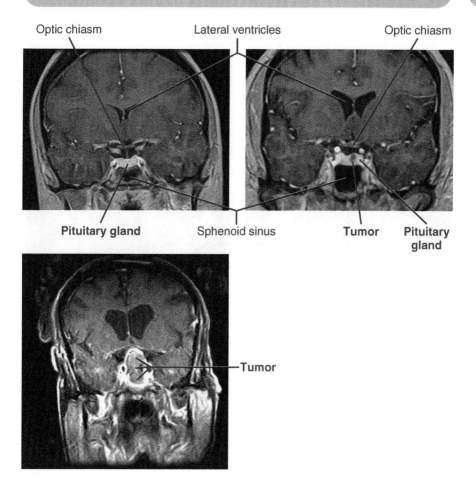

Optic chiasm | Lateral ventricles | Optic chiasm

Pituitary gland | Sphenoid sinus | Tumor | Pituitary gland

—Tumor

Coronal contrast-enhanced MR image of a patient with a pituitary microadenoma *(upper right)* compared to a patient with normal anatomy *(upper left)*. Lacking the blood-brain barrier, the normal pituitary enhances intensely (bright area), outlining pituitary microadenomas that are often relatively hypovascular (dark lesion within the pituitary). The upper right MR image was requested for a patient with hyperprolactinemia. The bottom MR image is for a patient who has a pituitary macroadenoma in which the tumor is clearly impinging on the optic chiasm, resulting in visual field deficits.

GAS: Cranial nerve lesions (p. 855)
COA: Visual field defects (p. 1080)

Pineal Gland Cyst

Corpus callosum

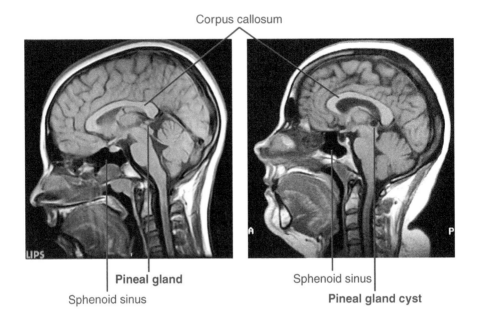

Pineal gland

Sphenoid sinus

Sphenoid sinus

Pineal gland cyst

Sagittal MR images, with the normal image on the left. The patient shown on the right has a benign pineal gland cyst. Before the development of MRI, pineal cysts were considered rare. Now, small (<5 mm) pineal cysts are considered to be common incidental benign findings. Larger cysts may be symptomatic, often associated with headaches in 40- to 49-year-old women.

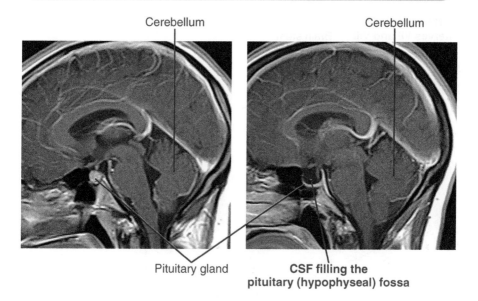

Cerebellum · Cerebellum

Pituitary gland

CSF filling the
pituitary (hypophyseal) fossa

Sagittal contrast-enhanced MR images. A brightly enhancing pituitary gland (hypophysis) with a normal nearly spherical shape is shown on the left. The patient on the right presented with headache and was found to have papilledema (a sign of increased intracranial pressure), raising clinical concern for brain tumor (see Primary Brain Tumor, page 4). The MR image showed the appearance of an "empty sella" (anatomically, an empty hypophyseal fossa) consistent with pseudotumor cerebri; the latter is related to a deficient diaphragma sellae, a benign syndrome associated with papilledema.

COA: Papilledema (p. 911)

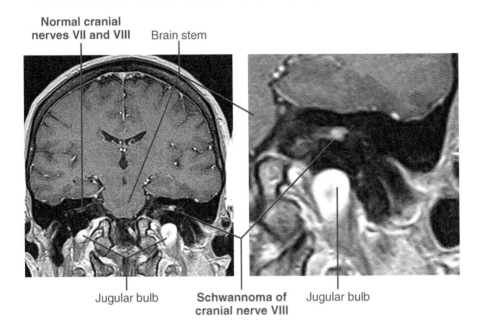

Normal cranial nerves VII and VIII — Brain stem

Jugular bulb — Schwannoma of cranial nerve VIII — Jugular bulb

Contrast-enhanced coronal head MR images (magnified view on the right). Whereas the patient's right cranial nerves VII and VIII appeared normal (right image), injection of IV contrast resulted in an enhancement (brightness) of a tumor of the left vestibulocochlear nerve (VIII) in the internal auditory canal (both images). This tumor can compress the facial nerve (VII). MRI is the procedure of choice for evaluation of sensorineural hearing loss.

GAS: Cranial nerve lesions (p. 855)
COA: Dizziness and hearing loss (p. 979)
COA: Injuries to vestibulocochlear nerve (p. 1082)

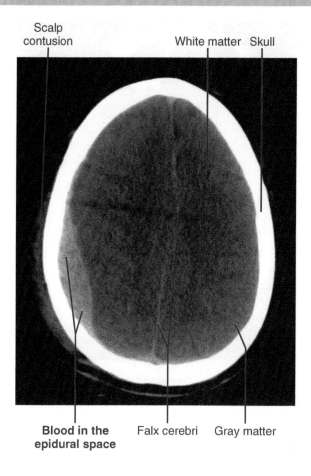

Scalp contusion

White matter Skull

Blood in the epidural space

Falx cerebri

Gray matter

Axial CT image of the head. The characteristic lens-like (biconvex) shape of the epidural hematoma is the result of the strong bond between the dura and the calvarium, especially at the sutures. An epidural hematoma usually results from a skull fracture that lacerates a branch of the middle meningeal artery. The rapid arterial bleeding causes significant hydrostatic pressure that indents the brain surface and exerts a "positive mass effect," which caused the obvious deviation of the falx cerebri toward the left. A posttraumatic collection of blood with this shape is treated as a surgical emergency.

GAS: Fractures of the skull vault and extradural hematoma (p. 829)
COA: Head injuries and intracranial hemorrhage (pp. 876-877)

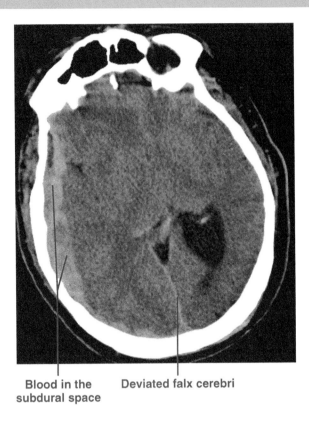

Blood in the subdural space　　**Deviated falx cerebri**

Axial CT image of the head. Note that acute hemorrhage is bright on CT. The subdural potential space is expanded during pathologic processes. In this case, veins have probably been lacerated, with progressive hemorrhage enlarging the subdural space over several hours or days. The hematoma causes pressure on the underlying brain, resulting in the deviation of the falx cerebri seen in the image. Serious neurologic injury or death may result if not treated. Initial treatment is to drill (trephine) a hole in the skull over the hematoma to relieve the pressure.

GAS: Types of intracranial hemorrhage (pp. 845-846)
COA: Head injuries and intracranial hemorrhage (pp. 876-877)

CSF filling the
subarachnoid space
within enlarged sulci
due to cerebral cortical
atrophy

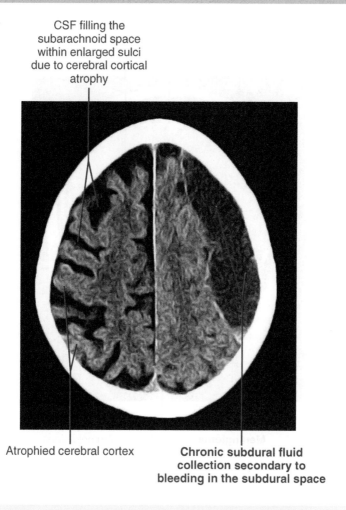

Atrophied cerebral cortex

**Chronic subdural fluid
collection secondary to
bleeding in the subdural space**

Axial CT image of the head of a patient with a chronic subdural hematoma on the left side. The subdural fluid collection can be identified as chronic because an acute hematoma would be much brighter (see p. 10). Subdural hematomas are usually of venous origin and progress slowly, as opposed to epidural hematomas that are of arterial origin and may reach maximum size within minutes. On the patient's right side, cortical atrophy results in diffusely thin gyri and widened sulci. Cortical atrophy increases the length of venous segments bridging the subdural space, making them susceptible to shear injury. In this patient, bilateral cortical atrophy predisposed the patient to developing a subdural hematoma. The mass effect of the chronic subdural hematoma has compressed underlying cortex and effaced the sulci on the patient's left.

GAS: Types of intracranial hemorrhage (pp. 845-846)
COA: Head injuries and intracranial hemorrhage (pp. 876-877)

Meningioma

Superior sagittal sinus

Lateral ventricle

"Dural tail"

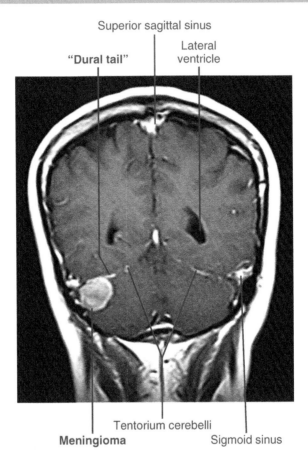

Tentorium cerebelli

Meningioma

Sigmoid sinus

Coronal contrast-enhanced MR image of a meningioma originating from the right side of the tentorium cerebelli. The tumor shows intense enhancement and the classic "dural tail" sign of a meningioma that provides evidence of its dural origin. Unenhanced CT may reveal characteristic calcification of meningiomas, many of which grow slowly or not at all after diagnosis, and therefore may not require surgery. These meningomas also enhance in contrast-enhanced CT studies.

GAS: Brain tumors (p. 835)

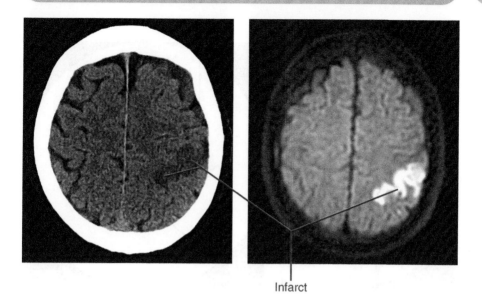

Infarct

Axial CT image *(left)* showing edema and diffusion-weighted axial MR image *(right)* showing restricted diffusion in the left parietal lobe (primary sensory area) following ischemic stroke involving obstruction of the middle cerebral artery. The restricted intracellular diffusion revealed by this MR scan is apparent within 1 hour of a stroke, whereas the extracellular edema shown in the CT image is not apparent for 6 hours to days after the event. The patient presented with abrupt onset of right-side facial, upper limb, and upper trunk numbness.

GAS: Stroke (p. 839)
COA: Strokes (pp. 887-888)

Aneurysm

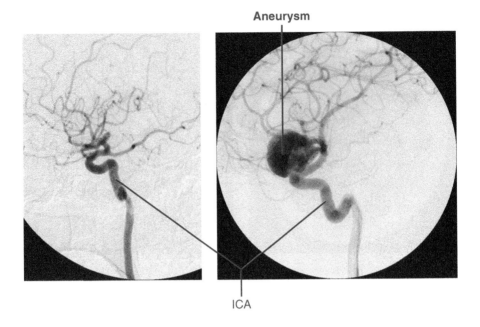

ICA

Angiograms of the internal carotid artery (ICA), with the normal image on the left. The angiogram of the patient on the right shows a large aneurysm of the distal ICA. These angiograms require passage of an intraarterial catheter for injection of radiographic contrast material. Intracranial aneurysms may be treated either by transcatheter techniques, in which a metal coil is placed in the aneurysm to cause thrombosis, or by surgery, during which a clip is placed across the neck of the aneurysm.

GAS: Intracerebral aneurysms (pp. 840-841)
COA: Strokes (pp. 887-888)

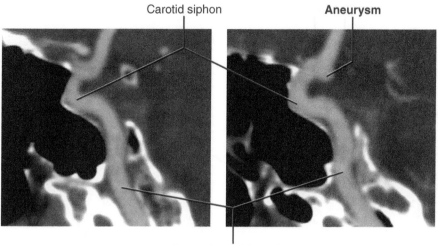

Carotid siphon   **Aneurysm**

Internal carotid arteries

Images from CT arteriography show both internal carotid arteries in one patient (the normal ICA is on the left). In the right image an aneurysm of the ICA just distal to the carotid siphon is evident as a bulging contour of one wall of the vessel. Very effective noninvasive MR arteriography is widely available to evaluate the intracranial circulation. However, if a patient is not a candidate for MRI (e.g., because of a pacemaker), CT angiography is a good alternative. Both of these imaging techniques are considered "non-invasive" because no intraarterial catheter is required.

GAS: Intracerebral aneurysms (pp. 840-841)
COA: Strokes (pp. 887-888)

Carotid Bifurcation Plaque

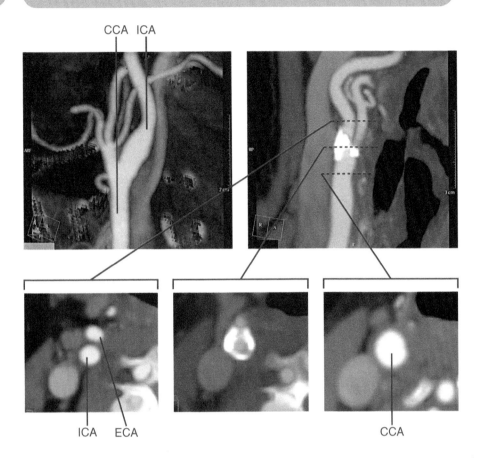

Ultrasonography is usually used for initial evaluation of a suspected carotid artery stenosis. Prior to carotid endartectomy or stenting, additional imaging is usually done with catheter, MR, or, as shown here, CT arteriography *(top left,* normal patient; *top right,* calcified carotid bifurcation plaque)*. Calcified arteriosclerotic plaque is very dense *(white)* on CT *(top right* and *bottom middle)*. Because calcified plaque may obscure the vessel lumen in some projections, cross-sectional images *(bottom)* become critical for accurate assessment of lumen stenosis caused by calcified plaque. (CCA, common carotid artery; ICA, internal carotid artery; ECA, external carotid artery).

GAS: Stroke (p. 839)
COA: Brain infarction (p. 888)

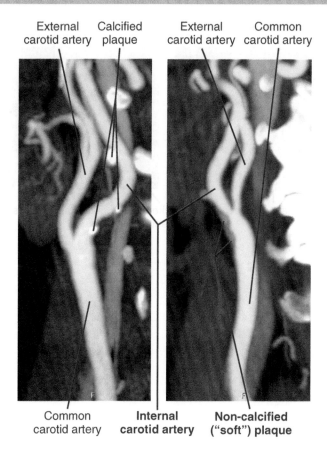

External carotid artery | Calcified plaque | External carotid artery | Common carotid artery

Common carotid artery | **Internal carotid artery** | **Non-calcified ("soft") plaque**

Left and right CT arteriograms from the same patient on p. 16. In the image on the left, bright calcified plaque of the internal carotid artery (ICA) causes no stenosis and is of little significance. On the contralateral side *(right)*, noncalcified or "soft" plaque (fatty and/or fibrous atheroma) at the carotid bifurcation causes a significant stenosis of the origins of both the external and internal carotid arteries. Severe ICA stenosis caused by calcified plaque (see p. 16) is usually treated surgically by endarterectomy. ICA stenosis caused by soft plaque may be treated by stenting.

GAS: Stroke (p. 839)
COA: Brain infarction (p. 888)

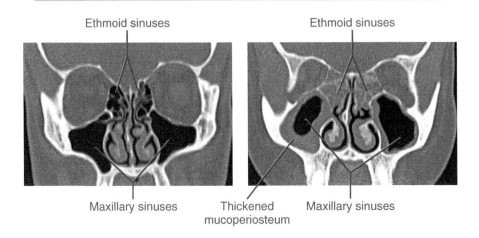

Ethmoid sinuses   Ethmoid sinuses

Maxillary sinuses   Thickened   Maxillary sinuses
                    mucoperiosteum

Coronal CT images of the face; in the patient with maxillary sinusitis *(right)*, note the greatly thickened mucoperiosteum lining the maxillary sinuses compared with the normal image on the left. The ethmoid sinuses in the image on the right are "completely opacified," a common radiologic phrase indicating that normal air spaces are completely filled with fluid and/or abnormal mucoperiosteal thickening.

COA: Sinusitis (p. 964)

Anterior fossa of
the cranial cavity

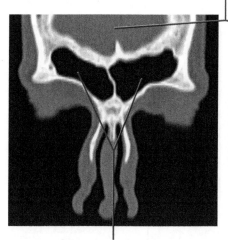 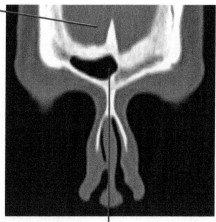

Aerated frontal sinuses   **Aerated right frontal sinus**

Coronal CT images of the head; frequent variations are seen in paranasal sinus pneumatization, with asymmetric sinus development quite common. This asymmetry is illustrated in the right image, in which the left frontal sinus never became aerated. Note that even in the "normal" image *(left)* the frontal sinuses are not perfectly symmetric.

Blow-Out Fractures

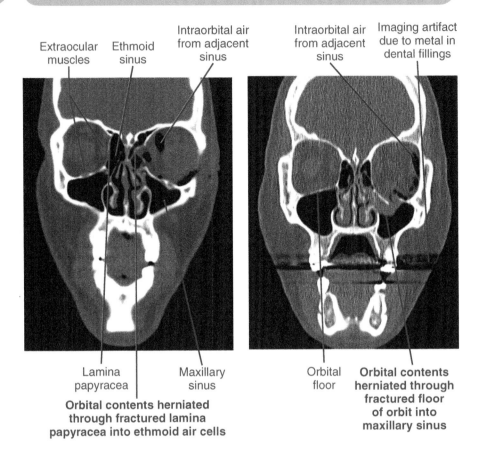

Extraocular muscles — Ethmoid sinus — Intraorbital air from adjacent sinus — Intraorbital air from adjacent sinus — Imaging artifact due to metal in dental fillings

Lamina papyracea — Maxillary sinus — Orbital floor — **Orbital contents herniated through fractured floor of orbit into maxillary sinus**

Orbital contents herniated through fractured lamina papyracea into ethmoid air cells

Coronal CT images showing blow-out fractures of the medial wall of the left orbit *(left)* and the floor of the left orbit *(right)*. Air from the sinuses has entered each orbit. This communication could allow bacterial contamination of the orbit from the sinus. If only intraorbital fat herniates through the fracture, ocular compromise is not likely. However, an extraocular muscle can become entrapped in the fracture, resulting in diplopia; this is an indication for surgical repair.

COA: Fractures of the orbit (p. 909)

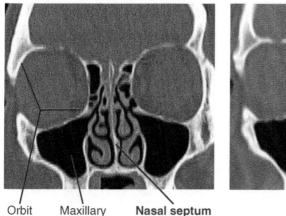 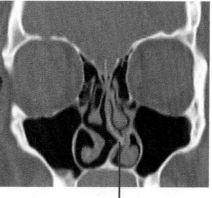

Orbit   Maxillary   **Nasal septum**   **Deviated nasal septum**
       sinus

Coronal CT images of the head; normally, the nasal septum approximates the midsagittal plane *(left)*. In cases where the septum is severely deviated to one side or the other, the patient may have difficulty breathing and/or recurrent sinusitis attributable to improper drainage of paranasal sinuses (in this case, likely the maxillary sinus).

COA: Deviation of nasal septum (p. 963)

Nasal Bone Fracture

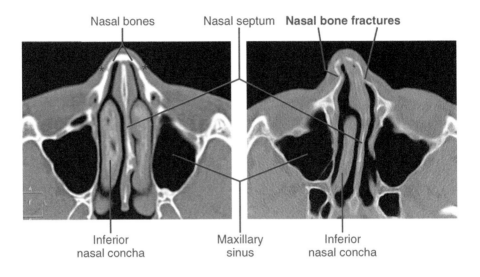

Nasal bones   Nasal septum **Nasal bone fractures**

Inferior
nasal concha   Maxillary
sinus   Inferior
nasal concha

Axial CT images, with the normal image on the left. The patient depicted in the right image sustained a comminuted nasal bone fracture. There is an incidental nasal septal deviation. In the patient with normal anatomy *(left)*, the symmetric osseous "gaps" *(asterisks)* represent the nasomaxillary sutures, which are distinguishable from a fracture because they are known normal features, are symmetric, and cause no deformity. If clinical examination indicates that the patient only suffered a nasal injury, then radiography is sufficient. However, if physical examination indicates that additional facial fractures may have occurred, then CT is the procedure of choice.

COA: Nasal fractures (p. 963)

Mandibular
condyle

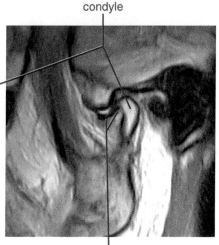
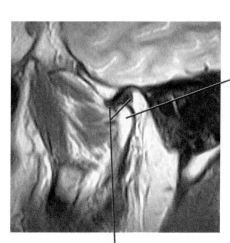

Articular disc

**Articular disc displaced
anterior to the condyle**

Sagittal MR images, with the normal image on the left. The articular disc (meniscus) appears as a dark, bow tie–shaped structure. In the image with the displaced articular disc *(right)*, the disc has been displaced anteriorly. Patients usually present with temporomandibular joint pain, a popping or clicking sensation, and limited mouth opening.

COA: Dislocation of TMJ (p. 927)

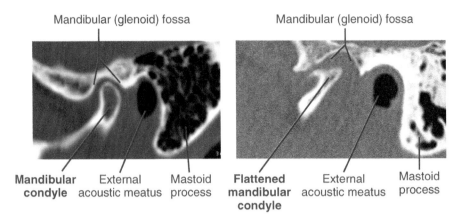

Mandibular (glenoid) fossa — Mandibular (glenoid) fossa

Mandibular condyle — External acoustic meatus — Mastoid process — **Flattened mandibular condyle** — External acoustic meatus — Mastoid process

Sagittal CT images; the normal image is on the left. The right image shows a flattened mandibular condyle resulting from degenerative arthropathy of the joint. MR imaging is needed for evaluation of the articular disc, but for ideal demonstration of arthritic osseous changes, CT scanning is commonly performed.

COA: Arthritis of TMJ (p. 927)

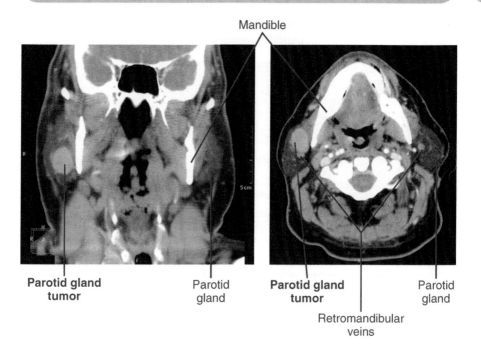

Mandible

Parotid gland tumor

Parotid gland

Parotid gland tumor

Parotid gland

Retromandibular veins

Coronal *(left)* and axial *(right)* CT images of a patient with a parotid mass. The most common parotid tumor is a benign pleomorphic adenoma, comprising approximately 80% of all salivary gland tumors. Malignancies also occur in the salivary glands. A risk from parotid surgery is potential injury to the facial nerve or one or more of its branches within the parotid gland. Cross-sectional imaging is valuable in surgical planning. Although these nerves cannot be seen by CT imaging, tumors superficial to the retromandibular vein may be resected with much less risk of facial nerve injury than tumors deep to the vein.

GAS: Parotid gland (p. 865)
COA: Parotidectomy (p. 926)

Calculus, at sublingual
papilla (opening of duct)

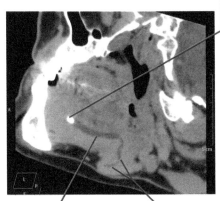 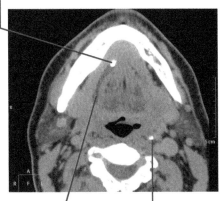

| Dilated submandibular duct in floor of mouth | **Dilated duct system within submandibular gland** | **Dilated submandibular duct in floor of mouth** | Styloid process |

Oblique sagittal *(left)* and axial *(right)* CT images. When sialolithiasis is clinically suspected, CT scanning is the most sensitive imaging procedure. Unenhanced scanning should be done so that small, brightly enhancing vessels are not confused with calculi, most of which are calcified and very bright on CT.

COA: Excision of submandibular gland and removal of a calculus (p. 950)

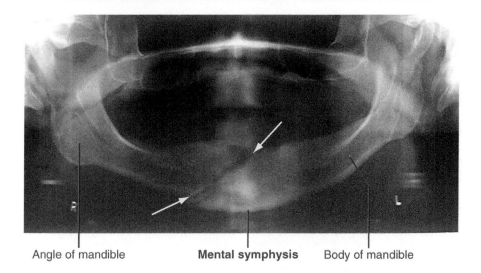

Angle of mandible   **Mental symphysis**   Body of mandible

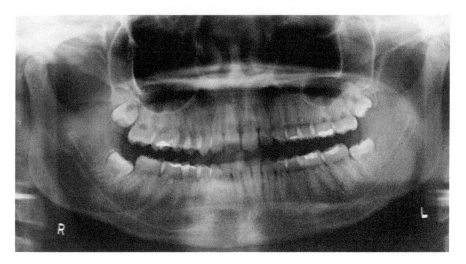

The top image is a panoramic radiograph showing a fracture of the mandible near the symphysis *(yellow arrows)*. Note that this is an edentulous patient. The lower image shows a similar panoramic view of a patient with normal anatomy. Dentists and oral surgeons commonly use these special radiographic projections of the mandible.

COA: Fractures of mandible (pp. 837-838)

Basal Skull Fracture

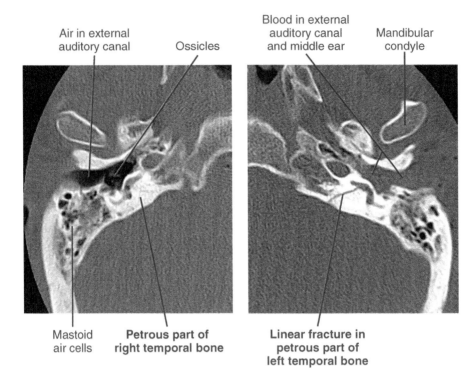

Air in external
auditory canal

Ossicles

Blood in external
auditory canal
and middle ear

Mandibular
condyle

Mastoid
air cells

**Petrous part of
right temporal bone**

**Linear fracture in
petrous part of
left temporal bone**

Axial high resolution thin section CT images of a patient who sustained a
linear skull base fracture on his left side. Routine "head CT" images may not
reveal such a fine skull base fracture. As a result of the injury, the external
auditory canal and middle ear are filled with blood instead of air.

COA: Fractures of cranial base (p. 876)

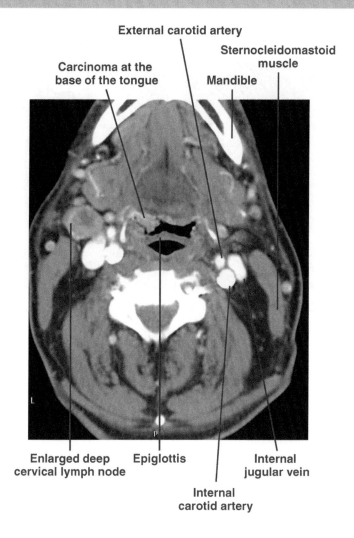

External carotid artery

Sternocleidomastoid muscle

Carcinoma at the base of the tongue

Mandible

Enlarged deep cervical lymph node

Epiglottis

Internal jugular vein

Internal carotid artery

Cervical lymphadenopathy is common in the presentation of cancers of the head and neck and indicates that such cancers have already metastasized at the time of diagnosis. In this axial CT image, note the bright lobulated lesion at the base of the right side of the tongue. In this nicotine user, it is nearly certain to be a mucosal cancer. The enlarged lymph node (with a necrotic low-density center that is common with this kind of metastatic disease) was palpable by the patient and was the "chief complaint" that led this patient's physician to request the CT scan.

GAS: Clinical lymphatic drainage of the head and neck (p. 985)
COA: Radical neck dissections (p. 1052)

Tongue (Lingual) Cancer

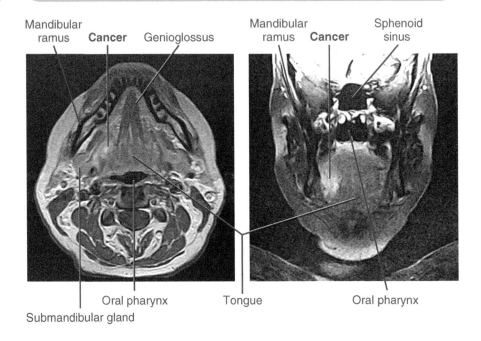

Axial *(left)* and coronal *(right)* contrast-enhanced MR images depicting a cancer of the inferior aspect of the tongue on the patient's right side. The cancer is hyperintense (bright) because it accumulates more of the IV contrast material than surrounding tissues. Note the absence of similar bright tissue on the contralateral side. These cancers almost always occur in tobacco users. In many pharyngeal cancers, the diagnosis is established by direct visualization and biopsy, with cross-sectional imaging used for staging and to determine if surgical resection is possible.

COA: Lingual carcinoma (p. 950)

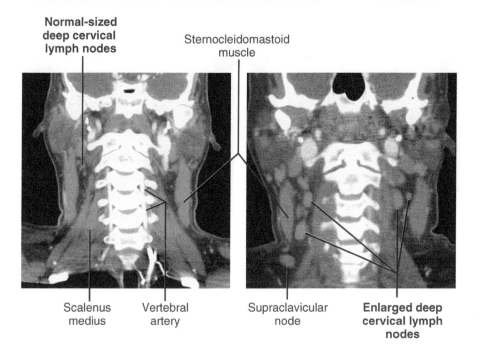

Normal-sized deep cervical lymph nodes

Sternocleidomastoid muscle

Scalenus medius

Vertebral artery

Supraclavicular node

Enlarged deep cervical lymph nodes

Coronal CT scans of the neck; in the patient with normal anatomy *(left)* the deep cervical lymph nodes are small. However, pathologically enlarged lymph nodes are easily seen in CT *(right)*. A supraclavicular node is sometimes referred to as a sentinel node because its enlargement is associated with malignant diseases of the thorax and abdomen, including lung cancer.

GAS: Clinical lymphatic drainage of the head and neck (p. 985)
COA: Radical neck dissections (p. 1052)
COA: Bronchogenic carcinoma (p. 125)

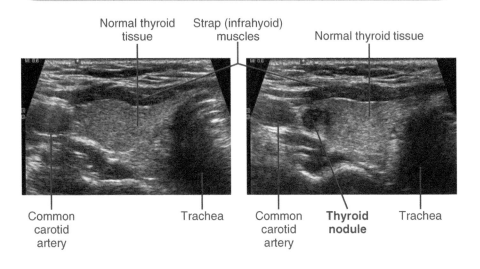

Normal thyroid tissue   Strap (infrahyoid) muscles   Normal thyroid tissue

Common carotid artery   Trachea   Common carotid artery   **Thyroid nodule**   Trachea

Ultrasound images from an examination of the thyroid gland. The image on the left shows a normal gland whereas the one on the right contains a nodule. The common carotid artery lies along the lateral margin of the thyroid gland and provides an important anatomic landmark. Anterior is at the top of both images.

GAS: Thyroid gland pathology (p. 968)

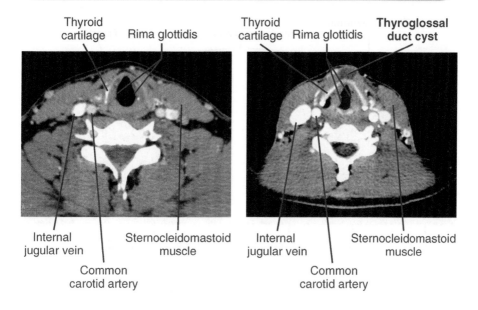

Axial CT images of the neck. The image on the right shows a thyroglossal duct cyst. No such fluid collection is found in the normal image *(left)*. Thyroglossal duct cysts develop from the epithelial remnants of the embryologic thyroglossal duct. They are also easily demonstrated by ultrasonography as a fluid collection in or near the midline anteriorly, most often between the isthmus of the thyroid gland and the hyoid bone.

GAS: Thyroid gland (p. 967)
COA: Thyroglossal duct cysts (p. 1041)

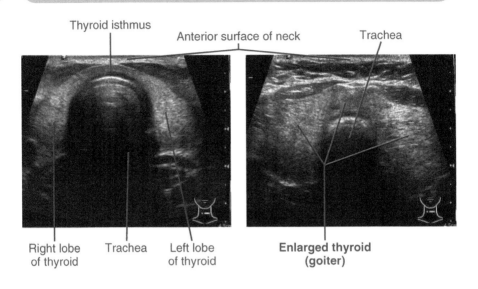

Thyroid isthmus

Anterior surface of neck

Trachea

Right lobe of thyroid · Trachea · Left lobe of thyroid

Enlarged thyroid (goiter)

Axial ultrasound images of the neck showing the thyroid gland and the trachea (normal image on the left). Different magnifications were used in the two images; the tracheas are actually about the same size, so the goiter shown (on the right) is quite large. Thyroid enlargement may be secondary to one or more thyroid nodules or, as in this case, medical disease that causes diffuse thyroid enlargement.

GAS: Thyroid gland pathology (p. 968)

COA: Enlargement of thyroid gland (pp. 1042-1043)

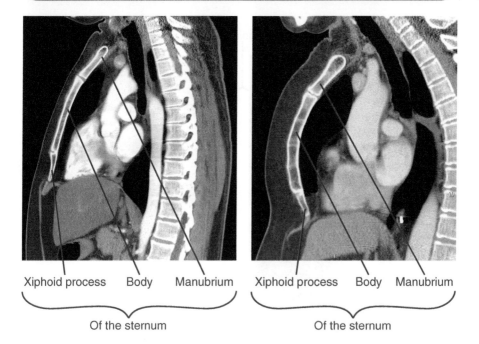

Xiphoid process　　Body　　Manubrium　　Xiphoid process　　Body　　Manubrium

Of the sternum　　　　　　　　　　　Of the sternum

Sagittal CT images of a normal chest *(left)* and the chest of a patient with pectus carinatum *(right)*. Pectus carinatum (also called "pigeon breast") is a protrusion abnormality of the chest wall that ranges from mild to severe. Whereas cosmetic concerns are important to some patients, others have dyspnea related to rigidity of the chest wall and decreased compliance of the lungs, or have an increased frequency of respiratory tract infections. There is also an association of this chest wall deformity with mitral valve prolapse. Pectus carinatum accounts for about 7% of developmental chest wall abnormalities.

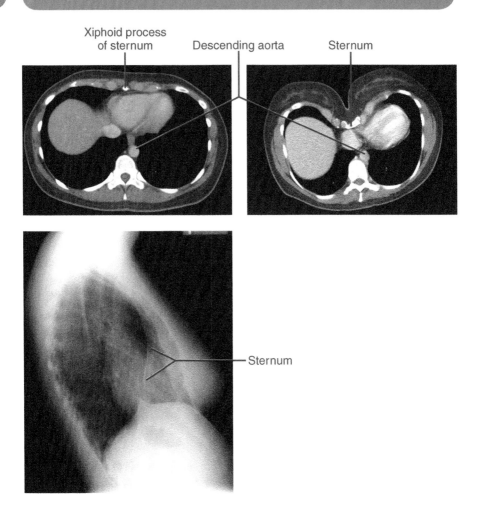

Axial CT images of a patient with a normal chest *(upper left)* and a patient with pectus excavatum *(upper right),* and a lateral radiograph from the latter patient *(bottom)*. Pectus excavatum is a developmental abnormality in which poorly coordinated osteocartilaginous growth of the anterior chest wall results in a "caved-in" appearance. In some patients the defect is less severe than that shown here and may only have psychosocial effects. In severe cases, cardiac and respiratory functions may be adversely affected. Pectus excavatum accounts for about 90% of developmental chest wall abnormalities.

Pneumothorax

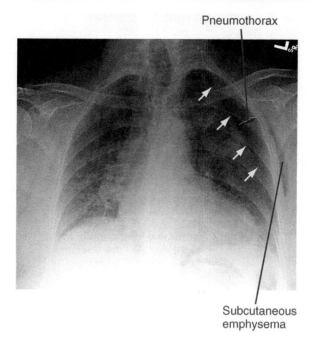

Subcutaneous emphysema

PA expiration chest radiograph. A radiograph taken during expiration shows a pneumothorax more conspicuously than during inspiration. Yellow arrows indicate the visceral pleural surface of the left lung. The pneumothorax is the dark collection of air between the lung surface and the chest wall. The dark stripe external to the chest wall is air; almost any collection of air or gas in soft tissues is referred to as subcutaneous emphysema. A radiograph of the chest during expiration exaggerates cardiac size and accounts for the increased density of the right lung. Some of the additional increased density in the middle of the left lung, in this case, is pulmonary contusion.

COA: Pneumothorax, hydrothorax, and hemothorax (p. 121)

Pneumonia

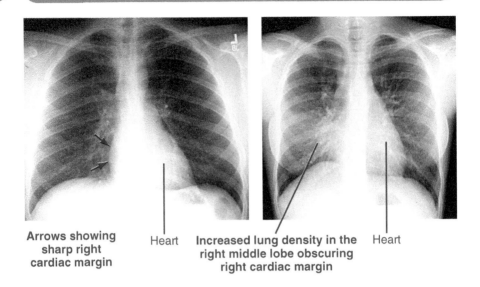

Arrows showing sharp right cardiac margin   Heart   **Increased lung density in the right middle lobe obscuring right cardiac margin**   Heart

PA chest radiographs, with the normal chest shown on the left. In the normal chest, the difference in radiographic density between the aerated lung and soft tissues results in sharp interfaces between the lung and adjacent structures. In the classic "silhouette sign," pneumonia increases the density of lung to that of soft tissue, and a normal margin, or edge, is obscured. The right radiograph shows a patient who has pneumonia in the medial segment of the right middle lobe that lies against the right heart border.

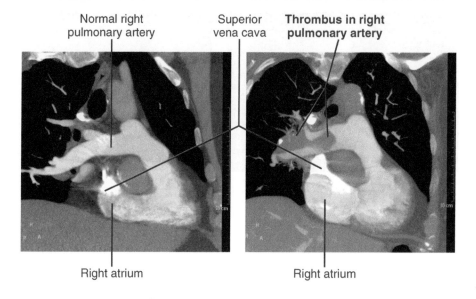

Normal right pulmonary artery | Superior vena cava | **Thrombus in right pulmonary artery**

Right atrium | Right atrium

Oblique coronal projections from a CT pulmonary arteriogram. IV contrast material is very bright in the visible portions of the superior vena cava and in the superior portions of the right atrium because the contrast material was injected into an antecubital vein. The contrast material mixes in the right atrium with darker blood (without contrast) from the inferior vena cava, resulting in moderate brightness in that chamber, the right ventricle, and the pulmonary artery. The relatively dark thrombus in the right pulmonary artery is clearly outlined by the brighter blood. CT is the best imaging procedure when pulmonary embolus is suspected.

COA: Pulmonary embolism (pp. 124-125)

Breast Cancer (Mammogram)

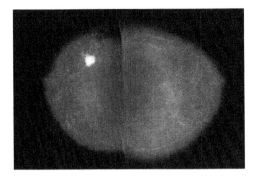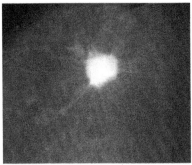

Craniocaudal mammography views (radiographic breast images) of a patient with a tumor in the right breast. The image on the right shows a high magnification view of the tumor, in this case a characteristically spiculated soft tissue mass that is easily seen in an otherwise fatty breast. Adipose tissue appears darker in mammograms than fibroglandular tissue or tumor. In a breast with little fat and a large amount of fibroglandular tissue, a tumor similar to the one shown above would be far less apparent.

GAS: Breast cancer (p. 139)
COA: Carcinoma of the breast (pp. 104-105)

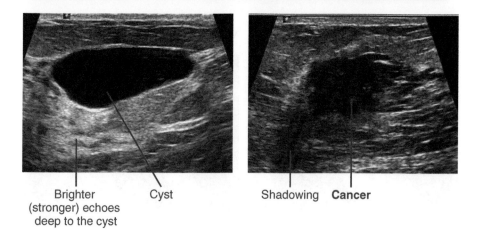

Brighter   Cyst   Shadowing **Cancer**
(stronger) echoes
deep to the cyst

Breast ultrasound examinations; the skin surface is at the top in both images. On the left, a simple benign cyst is diagnosed because it has no internal echoes, sharply defined smooth walls, and strong (bright) echoes deep to it. On the right, there is a mass that has irregular margins and decreased echoes (acoustic shadow) deep to it. This appearance is highly suspicious for breast cancer.

GAS: Breast cancer (p. 139)
COA: Carcinoma of the breast (pp. 104-105)

Mediastinal Tumor

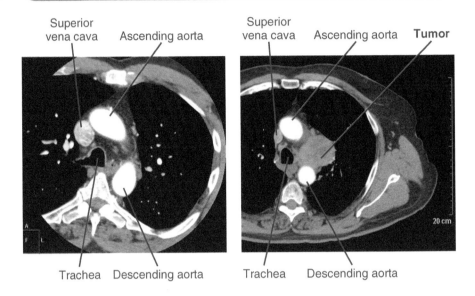

Axial CT images of a disease-free patient *(left)* compared with a patient who had a tumor in the left side of the superior mediastinum *(right)*. This tumor is possibly a lung malignancy that metastasized to the aorticopulmonary lymph node, which is found between the aortic arch superiorly and the pulmonary trunk inferiorly. The patient presented with hoarseness because the tumor compressed (or possibly invaded) the left recurrent laryngeal nerve.

GAS: The vagus nerves, recurrent laryngeal nerves, and hoarseness (p. 214)
COA: Injury to recurrent laryngeal nerves (p. 1043)

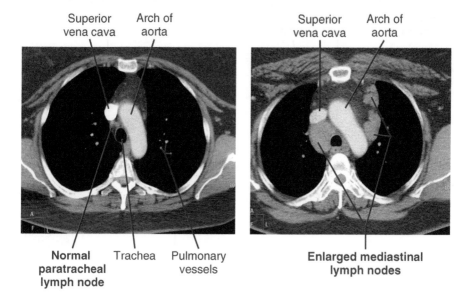

Superior vena cava   Arch of aorta

Superior vena cava   Arch of aorta

Normal paratracheal lymph node   Trachea   Pulmonary vessels

Enlarged mediastinal lymph nodes

Axial CT images; the image on the right is from a patient with lymphoma and depicts enlarged mediastinal lymph nodes (adenopathy). In the image on the left, a normal-sized paratracheal node is visible. For the patient with lymphoma, this CT scan had been recommended because radiography of the chest (for persistent cough) showed a widened mediastinum.

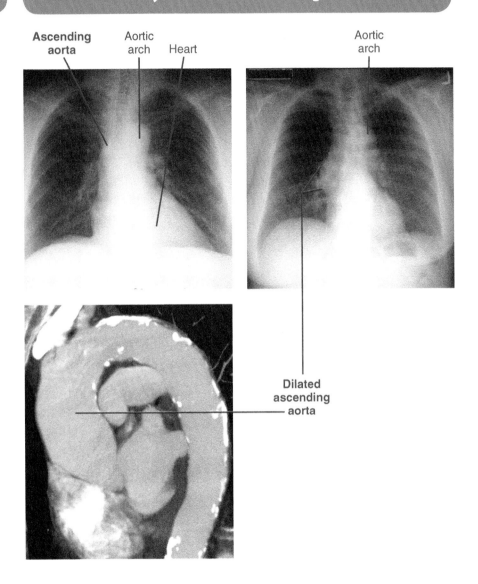

PA chest radiographs (*top*), with the normal image on the left. In patients younger than the one shown here the ascending aorta is small and near the midline. The right margin of the ascending aorta becomes more pronounced with the normal vascular changes associated with aging. Nevertheless, a subjectively "prominent" ascending aorta (*top right*) is suspicious for an aneurysm (dilated vessel with thinned walls), which can be confirmed by CT (*bottom*).

COA: Aneurysm of ascending aorta (p. 175)

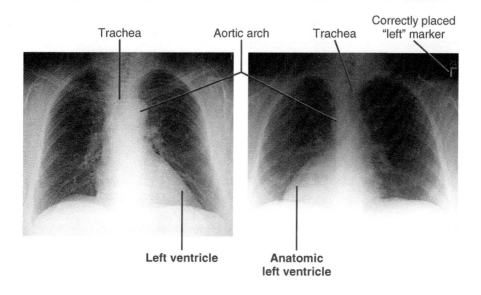

Trachea   Aortic arch   Trachea   Correctly placed "left" marker

Left ventricle   **Anatomic left ventricle**

PA chest radiographs; the normal image is on the left. The image on the right has not been printed incorrectly (note "L" marker). If the heart is on the right side of the chest but the abdominal organs are normally situated, then the condition shown in the right radiograph would be termed dextrocardia. Cardiac auscultation could be confusing in either situs inversus or dextrocardia. In total situs inversus, cholecystitis or appendicitis would cause left-sided symptoms that could confound initial medical evaluation. Approximately 25% of patients with situs inversus have dysfunctional cilia, leading to chronic sinusitis and bronchitis.

GAS: Aortic arch and its anomalies (p. 211)
COA: Positional abnormalities of the heart (p. 134)

Right Aortic Arch

Trachea   **Aortic arch**   **Aortic arch**   Trachea

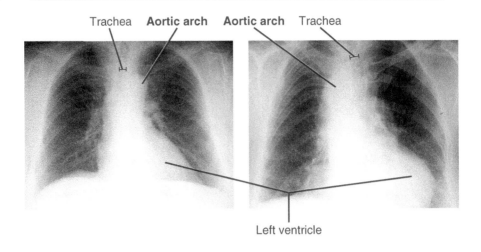

Left ventricle

PA chest radiographs (normal on the left). The right radiograph shows a patient who has a dilated right-sided aortic arch that is causing deviation of the trachea towards the left. The dilation of the arch may be atherosclerotic or be attributable to several other causes. There are a large variety of aortic anomalies that result in a right-sided arch. If the anomalous arch is a mirror image of the normal arch, there is a high probability of a major cardiac anomaly. A right-sided aortic arch may be asymptomatic if it is not associated with a vascular ring (arch and one or more great vessels encircling the trachea and/or esophagus; see Aberrant Right Subclavian Artery, p. 48).

GAS: Aortic arch and its anomalies (p. 211)
COA: Positional abnormalities of the heart (p. 134)

Superior vena cava | Trachea | Ascending aorta

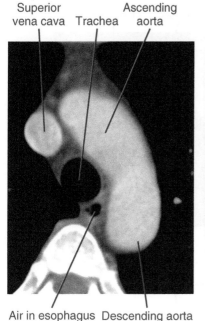

Superior vena cava | Ascending aorta | Trachea

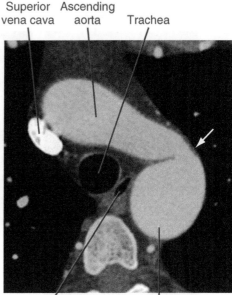

Air in esophagus | Descending aorta

Air in esophagus | Descending aorta

Axial CT images of the superior mediastinum (normal on left). The right image shows a patient with coarctation of the aorta, a congenital condition that usually is associated with the development of collateral circulation. The white arrow in the right image shows the site of the coarctation, which is typically near the attachment of the ligamentum arteriosum.

Image on right from: Chung JH, Gunn ML, Godwin JD, Takasugi J, Kanne JP: Congenital thoracic cardiovascular anomalies presenting in adulthood: A pictorial review, *Journal of Cardiovascular Computed Tomography* 3:535-536, 2008.

GAS: Coarctation of the aorta (p. 210)
COA: Coarctation of aorta (p. 175)

Aberrant Right Subclavian Artery

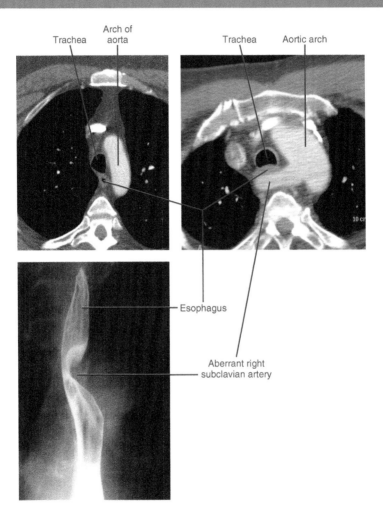

Axial CT of the superior mediastinum in two patients at the level of the aortic arch (*normal, top left*). The upper right image shows an aberrant origin and path of the right subclavian artery (as compared with normal, shown in the upper left image). The aberrant artery arises from the descending aorta, passes posterior to the esophagus, and then assumes its normal path. This vessel is not a brachiocephalic artery; additional CT sections showed the right common carotid artery arising independently from the aortic arch. The CT scan was recommended because a radiographic esophagram showed the characteristic deformity caused by a retroesophageal aberrant right subclavian artery (*bottom*).

GAS: Abnormal origin of great vessels (p. 211)
COA: Variations of great arteries (p. 174)

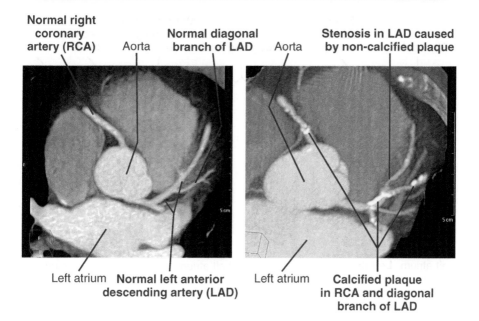

Normal right coronary artery (RCA) — Aorta — Normal diagonal branch of LAD — Aorta — Stenosis in LAD caused by non-calcified plaque

Left atrium — Normal left anterior descending artery (LAD) — Left atrium — Calcified plaque in RCA and diagonal branch of LAD

Coronary CT angiograms readily demonstrate constrictions (stenoses) of the coronary arteries associated with fatty atherosclerotic plaque as well as chronic calcified plaque. The risk of intraplaque hemorrhage, with acute occlusion of a coronary artery, is much greater for fatty plaque than for calcified plaque.

GAS: Coronary artery disease (p. 196)
COA: Coronary atherosclerosis (p. 156)

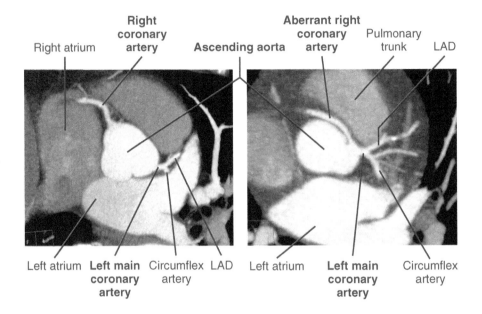

Right atrium — **Right coronary artery** — **Ascending aorta** — **Aberrant right coronary artery** — Pulmonary trunk — LAD

Left atrium — **Left main coronary artery** — Circumflex artery — LAD — Left atrium — **Left main coronary artery** — Circumflex artery

Axial CT images; on the right is an image of a patient whose right and left coronary ostia arise from the left aortic sinus. The CT image of a patient with normal anatomy is shown on the left. Note on the right that the right coronary artery could be compressed between the aorta and pulmonary trunk. Such anomalous coronary artery origins have caused fatal myocardial infarctions in young individuals.

LCX   LAD

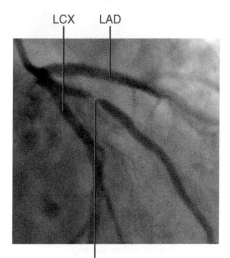

**Stenosis in a large
marginal branch of the LCX**

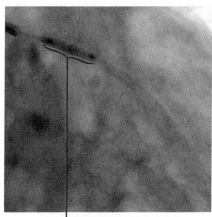

**Balloon inflated
within stenotic coronary
artery segment**

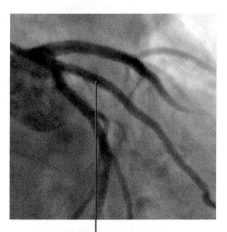

**Post angioplasty, lumen has
been restored**

Coronary arteriography *(above left)* shows stenosis of a large marginal branch of the left circumflex coronary artery (LCX). Other coronary arteries, including the left anterior descending (LAD), were normal. An angioplasty balloon was inflated within the stenosis *(above right)*. Repeat arteriography post angioplasty shows restoration of the normal coronary artery lumen *(bottom)*.

GAS: Coronary artery disease (p. 196)
COA: Coronary angioplasty (p. 157)

Aortic Valve Stenosis

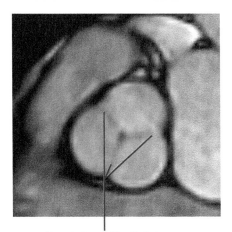

Coaptation of leaflets in
normal tricuspid aortic
valve during diastole

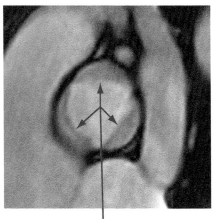

**Normal opening of the
aortic valve during
systole**

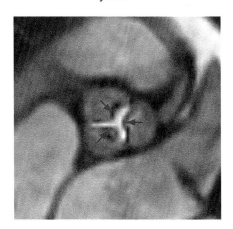

**Aortic valve stenosis: limited opening
of aortic valve during systole**

MR images of normal aortic valve, in diastole *(above left)* and systole *(above right)*, showing complete closing and opening of the valve. MR image of aortic valve stenosis in systole *(bottom)* shows how the thickened leaflets *(arrows)* fail to open adequately. The area of valve opening can be measured quantitatively to determine the hemodynamic significance of the stenosis. In these MR images flowing blood is bright.

Images provided by: Scott Mattson, DO, Medical Director, Echocardiology Laboratory, Cardiovascular MRI Laboratory, Lutheran Hospital of Indiana, Fort Wayne, IN.

GAS: Valve disease (p. 191)
COA: Aortic valve stenosis (p. 154)

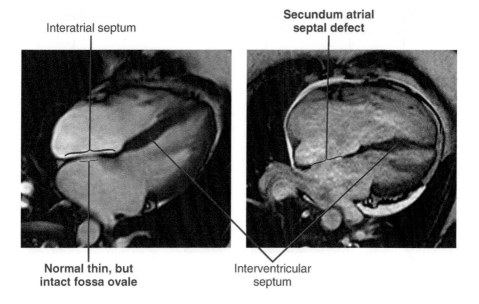

Images labeled:
- Interatrial septum
- Secundum atrial septal defect
- Normal thin, but intact fossa ovale
- Interventricular septum

Cardiac MR images of a patient with normal anatomy *(left)* and of a patient with secundum atrial septal defect (ASD) *(bottom)*. Approximately 15% of adults have a patent foramen ovale, usually of little significance. ASDs large enough to result in a significant left-to-right shunt are among the more common congenital cardiac defects diagnosed in adulthood. When diagnosed before age 25, repair of the defect usually results in a normal lifespan. The defect can be repaired either surgically or by using newer percutaneous techniques in which an umbrella-like device is passed via a catheter and expanded at the defect.

Images provided by: Scott Mattson, DO, Medical Director, Echocardiology Laboratory, Cardiovascular MRI Laboratory, Lutheran Hospital of Indiana, Fort Wayne, IN.

GAS: Common congenital heart defects (p. 197)

Hypertrophic Cardiomyopathy

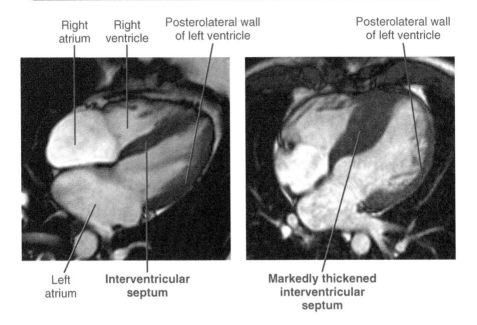

Right atrium | Right ventricle | Posterolateral wall of left ventricle | Posterolateral wall of left ventricle

Left atrium | **Interventricular septum** | **Markedly thickened interventricular septum**

Cardiac MR images, four-chamber view, at end diastole. In the patient with normal anatomy *(left)* there is symmetric muscular thickness of the interventricular septum and the posterolateral wall of the left ventricle. The image on the right shows a common presentation of hypertrophic cardiomyopathy in which the abnormal, thickened myocardium is mainly seen in the interventricular septum.

Images provided by: Scott Mattson, DO, Medical Director, Echocardiology Laboratory, Cardiovascular MRI Laboratory, Lutheran Hospital of Indiana, Fort Wayne, IN.

Internal mammary
(thoracic) artery

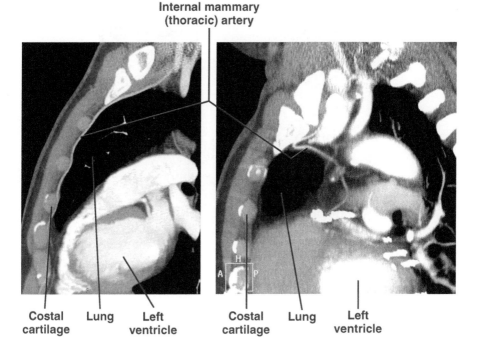

Costal cartilage   Lung   Left ventricle   Costal cartilage   Lung   Left ventricle

Sagittal CT images; anterior is to the left. In the image on the right, the internal mammary (thoracic) artery has been detached from the anterior thoracic wall for anastomosis to a coronary artery beyond an obstruction or severe stenosis, thus restoring myocardial blood flow. LIMA (left internal mammary artery) grafts are more common than RIMA (right internal mammary artery) grafts. Coronary bypass procedures also often use segments of the great saphenous vein or radial artery, anastomosed to the ascending aorta and to coronary arteries distal to a diseased coronary artery segment.

GAS: Coronary artery disease (p. 196)
COA: Coronary bypass graft (p. 156)

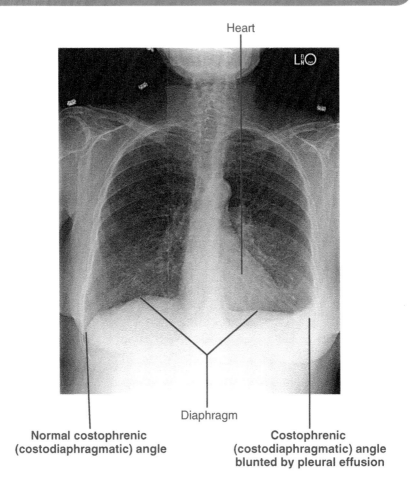

Heart

Diaphragm

**Normal costophrenic
(costodiaphragmatic) angle**

**Costophrenic
(costodiaphragmatic) angle
blunted by pleural effusion**

PA chest radiograph. In chest radiography, the normal costophrenic sulci are sharp, acute angles, as seen on the patient's right side in the image above. A pleural effusion will "blunt" this angle, resulting in a smooth, slightly concave superior margin as shown here on the patient's left. A similar appearance may be caused by postinflammatory scarring. When needed, a "decubitus" chest radiograph (done with the patient lying on his or her side) will show movement of fluid within the pleural space and differentiate effusion from scarring.

COA: Pneumothorax, hydrothorax, and hemothorax (p. 121)

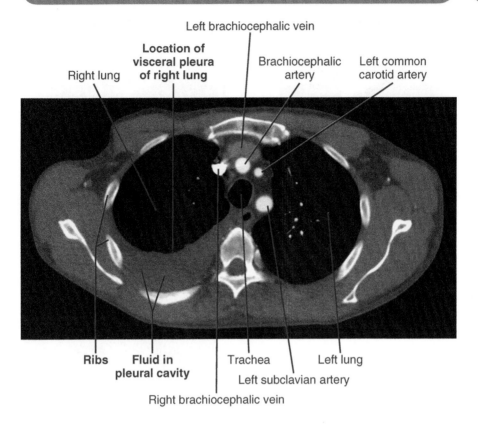

Left brachiocephalic vein

Location of visceral pleura of right lung

Right lung

Brachiocephalic artery

Left common carotid artery

Ribs

Fluid in pleural cavity

Trachea

Left subclavian artery

Left lung

Right brachiocephalic vein

Axial CT image of a right pleural effusion. The effusion is between the visceral and parietal layers of pleura (i.e., in the pleural cavity). Typically, as here, pleural layers are invisibly thin on CT images. On the patient's right side, the visceral pleura is at the interface of the dark lung parenchyma with the grey fluid. The parietal pleura on the right and both layers on the left are at the inner margin of the rib cage. On the patient's normal left side, the pleural cavity contains just enough fluid to lubricate movement between the layers of pleura during breathing. The patient was supine when scanned, so the effusion collected in the posterior right pleural cavity. The right brachiocephalic vein is very bright because a right arm vein was used for the IV contrast material injection; the left brachiocephalic vein is the density of blood without such contrast enhancement.

COA: Pneumothorax, hydrothorax, and hemothorax (p. 121)

Emphysema

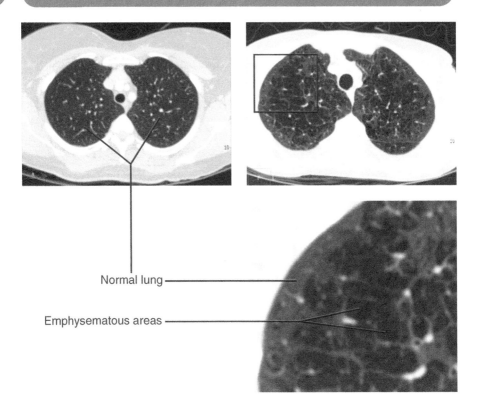

Normal lung

Emphysematous areas

CT images of a patient with normal lung tissue *(left)* and of a patient with emphysema *(upper right;* the lower right image is a magnified view of the indicated section). Note that normal lung tissue appears dark gray; emphysematous lungs have areas that appear black on CT because they are literally empty spaces (i.e., large, nonfunctional air sacs). Emphysema is the common end result of the airway damage caused by tobacco smoke.

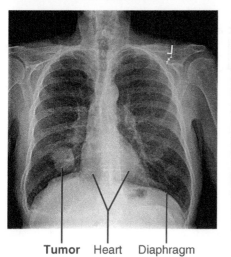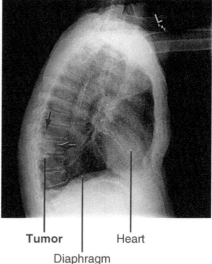

Tumor   Heart   Diaphragm   **Tumor**   Heart

Diaphragm

PA *(left)* and lateral *(right)* plain radiographs show a mass in the right lower lobe (tumor labeled and demarcated with red arrows). Needle biopsy established a diagnosis of primary lung cancer. From these images, tumor staging is not evident. Additional procedures will be needed to determine if the cancer has metastasized to hilar or mediastinal lymph nodes, or if distant metastases have occurred. See more advanced lung cancer case on p. 60.

GAS: Lung cancer (p. 175)
COA: Bronchogenic carcinoma (p. 125)

Lung Cancer, Advanced

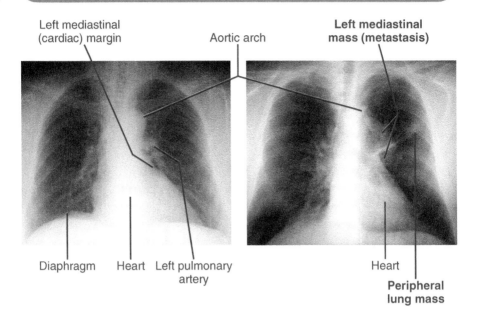

Left mediastinal (cardiac) margin

Aortic arch

Left mediastinal mass (metastasis)

Diaphragm   Heart   Left pulmonary artery

Heart

Peripheral lung mass

PA chest radiographs; the normal image is on the left. The left mediastinal border is normally concave just inferior to the aortic arch. In the lung cancer case on the right, a small peripheral lung mass has metastasized to the left side of the mediastinum and "fills in" that concavity. This obscures the left hilar structures, including the pulmonary artery. It is not uncommon for metastatic disease to be bulkier than the primary tumor. Because the PA image to the right is that of a nicotine user, advanced stage lung cancer was immediately suspected.

GAS: Lung cancer (p. 175)
COA: Bronchogenic carcinoma (p. 125)

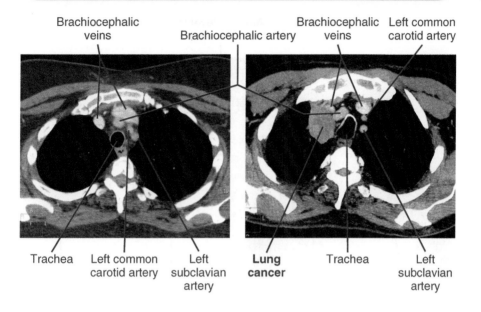

Brachiocephalic veins — Brachiocephalic artery — Brachiocephalic veins — Left common carotid artery

Trachea — Left common carotid artery — Left subclavian artery — **Lung cancer** — Trachea — Left subclavian artery

Contrast-enhanced axial CT images of a disease-free patient *(left)* and a patient with lung cancer *(right)*. The tumor evident in the right image was proven to be lung cancer by needle biopsy. The mass is compressing the right brachiocephalic vein. Because veins have thinner walls and lower intraluminal pressure than arteries, it is far more common for tumors to cause venous compression than arterial compromise. Edema and/or venous distention is occasionally the clinical sign that leads to suspicion of a tumor.

GAS: Lung cancer (p. 175)
COA: Bronchiogenic carcinoma (p. 125)

Large Sliding Hiatal Hernia

Air-fluid level in the stomach

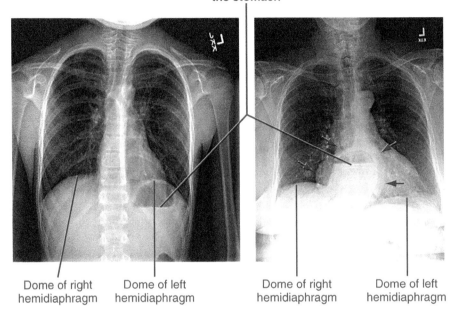

Dome of right hemidiaphragm | Dome of left hemidiaphragm | Dome of right hemidiaphragm | Dome of left hemidiaphragm

PA radiographs of the chest. The short red arrows in the image on the right outline the fundus and most of the body of the stomach, which are above the diaphragm (i.e., a sliding hiatal hernia). The normal image on the left shows the air-fluid level within the stomach in its usual position, inferior to the left hemidiaphragm. Whereas a very large sliding hiatal hernia can often be appreciated on a chest radiograph, if only a small portion of the stomach herniates through the diaphragm, endoscopy or barium esophagram may be required for diagnosis.

GAS: Hiatus hernia (p. 355)
COA: Hiatal hernia (p. 254)

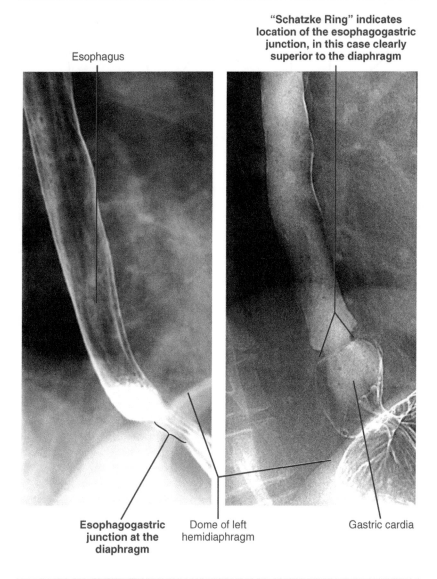

Esophagus

"Schatzke Ring" indicates location of the esophagogastric junction, in this case clearly superior to the diaphragm

Esophagogastric junction at the diaphragm

Dome of left hemidiaphragm

Gastric cardia

Air-contrast esophagrams. The normal image on the left shows the esophagogastric junction at the diaphragm. In the image on the right, the cardia of the stomach has "slid" through a widened hiatus and is superior to the diaphragm. A hiatus hernia often interferes with the function of the high-pressure zone that acts as a lower esophageal sphincter, allowing stomach acid to reflux into the esophagus (gastroesophageal reflux disease [GERD]).

GAS: Hiatus hernia (p. 355)
COA: Hiatal hernia (p. 254)

Esophageal Varices

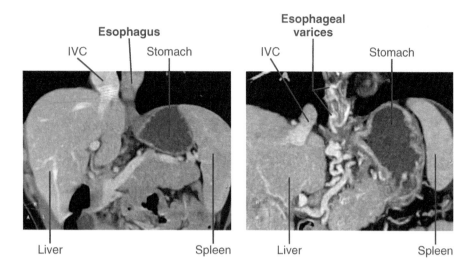

Esophagus

IVC   Stomach

Liver   Spleen

Esophageal varices

IVC   Stomach

Liver   Spleen

Coronal CT images; the patient with normal anatomy is shown in the left image. The right image shows a patient who has portal hypertension, resulting in esophageal varices (i.e., greatly distended submucosal veins). As portal vein pressure increases in cirrhosis, blood flow is reversed in the left gastric vein that normally drains those submucosal veins. The varices provide a shunt of blood from the portal system to systemic venous return via the azygos vein. Esophageal varices may become large and fragile, leading to catastrophic hemorrhage.

COA: Esophageal varices (p. 254)

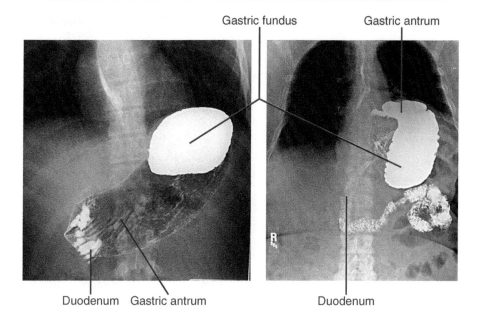

Gastric fundus   Gastric antrum

Duodenum   Gastric antrum   Duodenum

AP radiographs (normal image on left) from upper gastrointestinal (GI) radiographic examinations. Swallowed barium suspension is the white liquid in the GI lumen. An image from a patient with a large diaphragmatic hernia is presented on the right. In this patient, who had difficulty breathing, most of the stomach herniated through a diaphragmatic defect; however, because the gastroesophageal junction remained in its normal position within the esophageal hiatus of the diaphragm, the stomach rotated 180 degrees on its long axis, and is literally upside down.

GAS: Diaphragmatic hernias (p. 354)
COA: Rupture of diaphragm and herniation of viscera (p. 317)

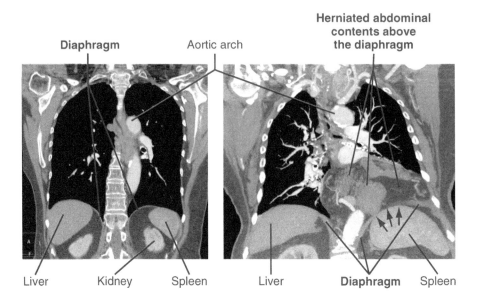

Coronal CT images (normal image on the left). The right image shows a patient who has a diaphragmatic hernia. The herniated structures include the stomach and adjacent intra-abdominal fat. Note the defect in the left hemidiaphragm *(arrows)*. The entire normal diaphragm may not be visible if it is the same CT density as adjacent tissue (e.g., along the superior margin of the liver). The herniation of the abdominal contents resulted in an abnormal orientation of the spleen (compare with left image of patient with normal anatomy).

GAS: Diaphragmatic hernia (p. 354)

COA: Rupture of diaphragm and herniation of viscera (p. 317)

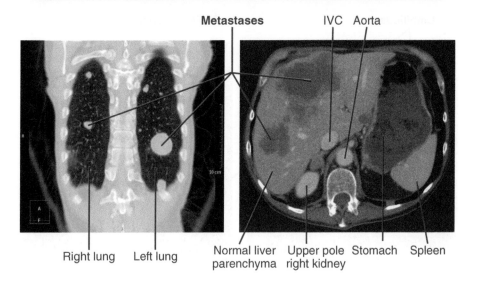

Coronal CT image of the thorax *(left)* and axial CT image of the upper abdomen *(right)*. When imaging reveals multiple tumors in an organ or system, such as shown in the lungs and liver above, it is reasonable to assume that these represent multiple metastases. This patient with metastatic colon carcinoma had lesions in the lungs and liver. On page 3 multiple metastatic brain lesions in a patient with breast cancer are shown. On page 162 a radionuclide bone scan showing extensive skeletal metastatic lesions is shown.

GAS: Breast cancer (staging the tumor; p. 139)
COA: The spread of cancer (p. 45)

Umbilical Hernia

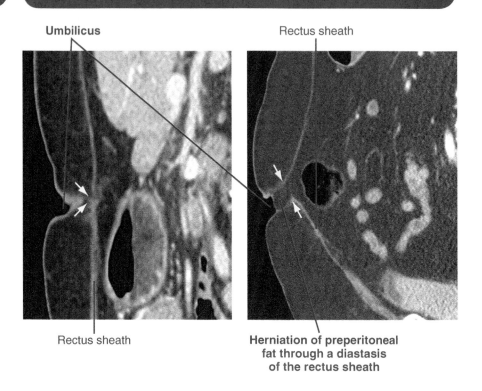

Umbilicus

Rectus sheath

Rectus sheath

Herniation of preperitoneal fat through a diastasis of the rectus sheath

Midsagittal CT images; on the right, an umbilical hernia is highlighted by the yellow arrows, which show a defect in the anterior abdominal wall and herniation of preperitoneal fat through a diastasis of the rectus sheath at the umbilicus. In the left image, the yellows arrows show an intact (normal) umbilicus.

GAS: Umbilical hernias (p. 291)
COA: Abdominal hernias (p. 197)

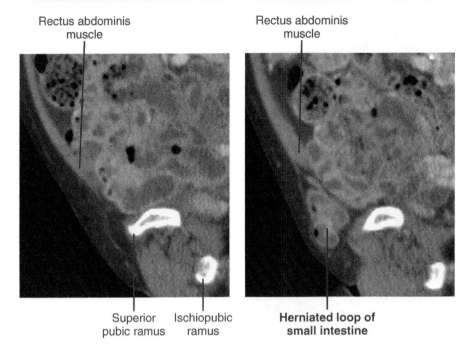

Rectus abdominis muscle

Rectus abdominis muscle

Superior pubic ramus | Ischiopubic ramus

Herniated loop of small intestine

Sagittal CT images of the left and right inguinal regions of the same patient. Note the inguinal hernia in the right image. The rectus abdominis muscle is the medial border of the inguinal triangle (of Hesselbach) through which a direct inguinal hernia protrudes.

GAS: Masses around the groin/inguinal hernias (pp. 290-291)
COA: Inguinal hernias (pp. 212-213)

Caput Medusae

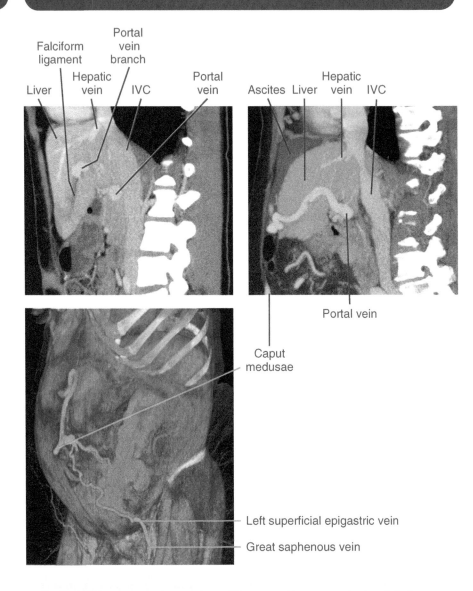

Sagittal CT images are shown above. The normal image on the top left shows only dark fat within the falciform ligament. In the patient with cirrhosis, shown on the top right, portal hypertension has resulted in a dilated paraumbilical vein within the falciform ligament that communicates with periumbilical superficial varicosities, forming a caput medusae. The 3D reconstruction (bottom) demonstrates anastomosis of the caput medusae with the great saphenous vein via the left superficial epigastric vein.

GAS: Hepatic cirrhosis/portosystemic anastomosis (pp. 339-340)
COA: Portal hypertension (p. 288)

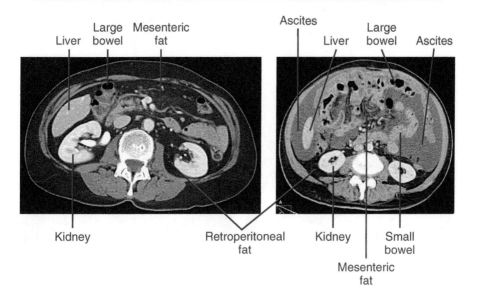

Axial CT scans of the abdomen. In the patient without ascites *(left)*, there is no visible intraperitoneal fluid. The image from the patient with ascites *(right)* shows considerable fluid in the peritoneal cavity. Bowel loops seem to be floating in the fluid. It is typical that the ascites displaces bowel from the flanks towards the midline. Note that the ascites within the peritoneal cavity helps to define the boundary between the peritoneal cavity and the retroperitoneal spaces.

GAS: Hepatic cirrhosis (p. 339)
COA: Peritonitis and ascites (pp. 223-224)

Abdominal Adenopathy

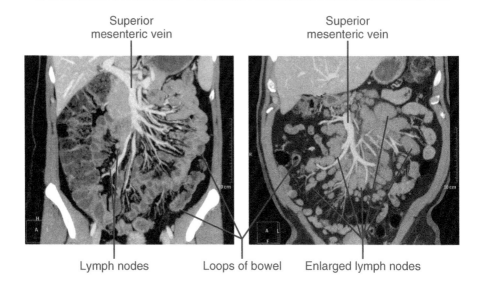

Superior mesenteric vein   Superior mesenteric vein

Lymph nodes   Loops of bowel   Enlarged lymph nodes

CT coronal images of the abdomen. Loops of bowel appear hollow and often cylindrical, whereas lymph nodes appear solid and ovoid or round. In the image on the left, some normal lymph nodes are seen along the ileocolic vein. In the image on the right, several of many enlarged lymph nodes are indicated. In a case of adenopathy such as this, lymphoma is suspected.

GAS: Retroperitoneal lymph node surgery (p. 373)
COA: Lymphangitis, lymphadenitis, and lymphedema (p. 46)

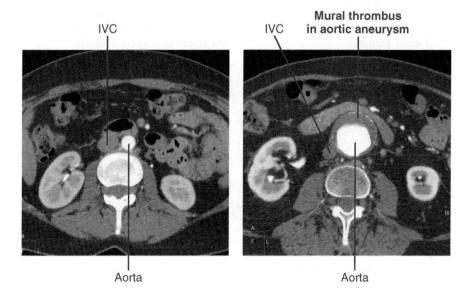

Axial CT arteriograms; the aortic lumen is opacified in these CT arteriograms by the rapid IV injection of iodinated contrast material. Many aneurysms contain a mural thrombus, as in the example on the right. Although ultrasound images can reveal aortic aneurysms, CT or MR scans are done for a more comprehensive analysis before treatment by stenting or surgery.

GAS: Abdominal aortic stent graft (p. 369)
COA: Pulsations of aorta and abdominal aortic aneurysm (p. 319)

Psoas Abscess

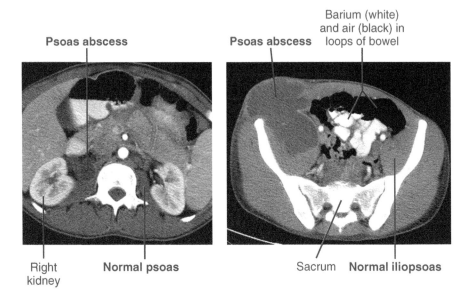

Psoas abscess   Psoas abscess   Barium (white) and air (black) in loops of bowel

Right kidney   Normal psoas   Sacrum   Normal iliopsoas

Axial CT images at different levels in a patient with a psoas abscess resulting from a vertebral column infection. The infection shown in both images extended from an infected vertebral body at the level of the kidneys into the adjacent right psoas muscle *(left image)*. From this level the infection infiltrated the psoas muscle and its fascia distally to the pelvis *(right image)*, and further distally into the upper thigh. Such a psoas abscess illustrates how pathologic processes often follow an anatomic structure or "pathway" throughout its course.

GAS: Psoas muscle abscess (p. 353)
COA: Psoas abscess (p. 318)

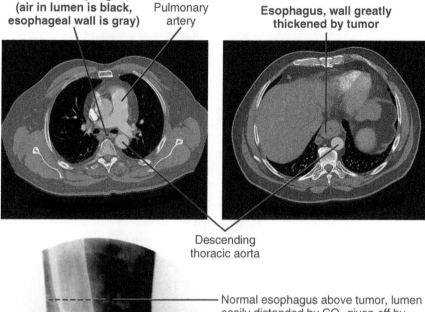

Normal esophagus (air in lumen is black, esophageal wall is gray)

Pulmonary artery

Esophagus, wall greatly thickened by tumor

Descending thoracic aorta

Normal esophagus above tumor, lumen easily distended by CO_2 given off by crystals swallowed by the patient during the upper GI examination

Narrowed and irregular lumen of esophagus, constricted by surrounding tumor

The bottom image is an upper GI radiograph of a patient with a gastroesophageal junction tumor and a hiatal hernia in which the two superimposed red dashed lines correspond to the levels of the top left and right axial CT images, respectively. The top right CT shows the markedly thickened esophageal walls associated with infiltration of the tumor.

GAS: Epithelial transition between abdominal esophagus and stomach (p. 303)

Duodenal Ulcer

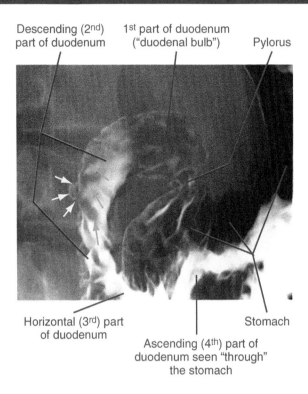

Descending (2nd) part of duodenum

1st part of duodenum ("duodenal bulb")

Pylorus

Horizontal (3rd) part of duodenum

Ascending (4th) part of duodenum seen "through" the stomach

Stomach

Radiograph from upper GI examination; thickened duodenal mucosal folds *(blue arrows)* "radiate" towards the duodenal ulcer *(yellow arrows)* that is a "crater" outside of the normal duodenal lumen. Most duodenal ulcers are in the first part of the duodenum ("duodenal bulb"). Postbulbar ulcers, as in this · patient, are suspicious for Zollinger-Ellison syndrome (hypergastrinemia, usually caused by an islet cell tumor of the pancreas).

GAS: Duodenal ulceration (p. 303)

COA: Duodenal ulcers (p. 257)

Ileal diverticulum

Terminal ileum

Transverse colon

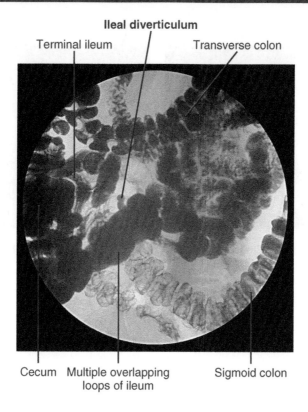

Cecum   Multiple overlapping
loops of ileum

Sigmoid colon

Radiograph (negative image) from fluoroscopy of a barium small-bowel study; the depicted diverticulum is in the typical location for a Meckel's diverticulum, 30-60 cm from the ileocecal junction. The relationships between bowel loops in these images can be very confusing because of their mobility and variability in position.

GAS: Meckel's diverticulum (p. 306)
COA: Ileal diverticulum (p. 258)

Hepatic Cirrhosis

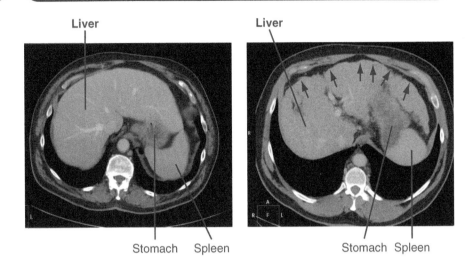

Liver   Liver

Stomach   Spleen   Stomach   Spleen

Axial CT images, normal image on the left. The right image shows a patient with cirrhosis of the liver (caused by hepatitis). Note the lobulated margin of the liver *(short red arrows)*, which is one of the characteristics of a cirrhotic liver. These lobulations may also be seen in gross specimens.

GAS: Hepatic cirrhosis (p. 339)
COA: Cirrhosis of the liver (p. 285)

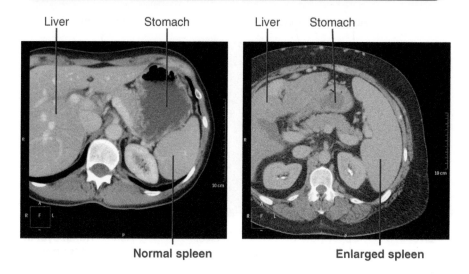

Liver   Stomach   Liver   Stomach

Normal spleen   **Enlarged spleen**

Axial CT scans; the moderately enlarged spleen in the right image displaces the stomach towards the midline (compare to the normal patient on the left). Ultrasound can accurately measure splenic size, but CT scanning offers a more comprehensive abdominal evaluation, especially with respect to detecting adenopathy associated with diseases that may cause splenomegaly.

GAS: Spleen disorders (p. 327)
COA: Splenectomy and splenomegaly (p. 281)

Renal Cyst (Simple)

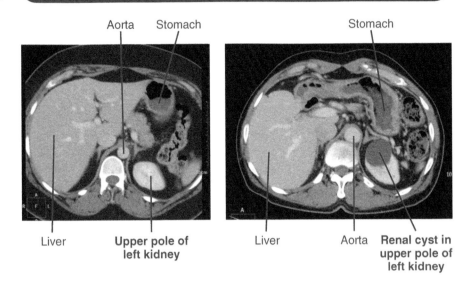

Aorta   Stomach   Stomach

Liver   Upper pole of   Liver   Aorta   Renal cyst in
          left kidney   upper pole of
                                                    left kidney

Axial CT images, normal image on left. A large renal cyst is apparent in the image on the right. "Simple" renal cysts, as the one shown here, have imperceptibly thin walls without septations, and are always benign. These are often discovered as incidental findings that increase in frequency with age. Renal cysts are also well visualized in abdominal ultrasound images. Cystic renal lesions are seen with a spectrum of complexity, from the simple cyst shown here to more complex cystic lesions with septations and thicker walls. The risk of malignancy increases progressively with such findings (see p. 81).

COA: Renal cysts (p. 298)

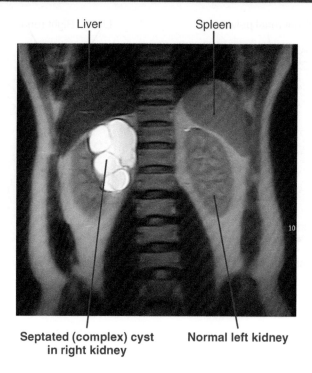

Liver   Spleen

Septated (complex) cyst
in right kidney   Normal left kidney

Coronal MR image of a patient with a septated (complex) cyst in the right kidney and a normal left kidney. Although the simple cyst depicted on p. 80 is very common and requires no follow-up, the risk of malignancy with more complicated cysts necessitates regular imaging surveillance.

COA: Renal cysts (p. 298)

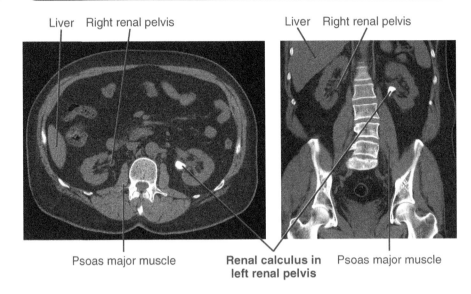

Liver   Right renal pelvis

Liver   Right renal pelvis

Psoas major muscle

Renal calculus in left renal pelvis   Psoas major muscle

Axial *(left)* and coronal *(right)* CT images of the abdomen depicting a calculus (stone) in the pelvis of the left renal collecting system. The vast majority of urinary tract "stones" are calcified and are visible on CT.

GAS: Urinary tract stones (p. 361)
COA: Renal and ureteric calculi (p. 300)

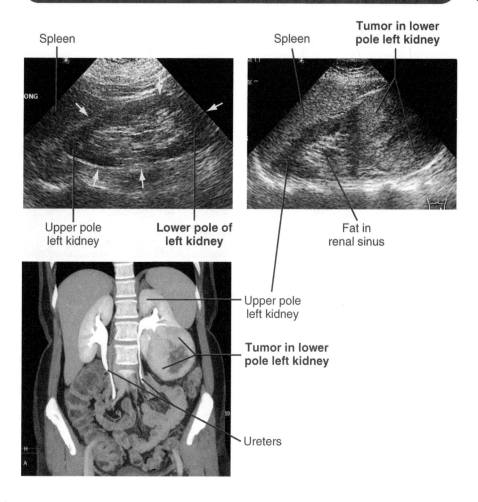

Spleen

Tumor in lower pole left kidney

Spleen

Upper pole left kidney

Lower pole of left kidney

Fat in renal sinus

Upper pole left kidney

Tumor in lower pole left kidney

Ureters

Ultrasound images of a normal left kidney *(upper left, yellow arrows)* and a left kidney with a large tumor in its lower pole *(upper right)*. A contrast-enhanced coronal CT image of the abdomen of the patient with the renal mass *(bottom)* is also shown. The tumor was initially discovered as an incidental finding during an abdominal ultrasound examination for abdominal pain. The CT was then requested for staging before surgery. The tumor is hyperechoic to normal renal cortex on the ultrasound examination and shows different CT density than normal renal parenchyma. On both examinations, the tumor greatly alters the normal renal contour.

GAS: Urinary tract cancer (pp. 361-362)

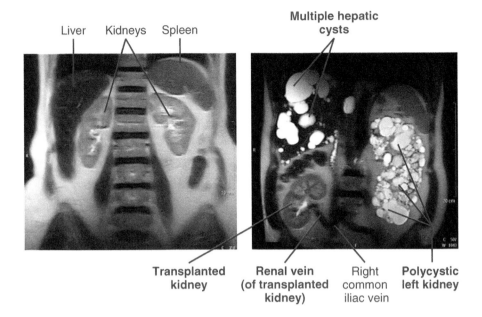

Liver   Kidneys   Spleen

Multiple hepatic cysts

Transplanted kidney   **Renal vein (of transplanted kidney)**   Right common iliac vein   **Polycystic left kidney**

Coronal MR images; the normal image is on the left. The right image shows a patient who has adult polycystic kidney disease. This patient has had the right kidney removed and has a transplanted kidney in the right iliac fossa. The vein of the transplanted kidney is anastomosed to the patient's right common iliac vein (in this MR image, flowing blood in vessels is dark). There are many cysts in the left kidney (static fluid is bright), with essentially no residual renal paren-chyma. Note the associated cysts in the liver.

GAS: Kidney transplant (p. 364)
COA: Renal cysts (p. 298)
COA: Renal transplantation (p. 298)

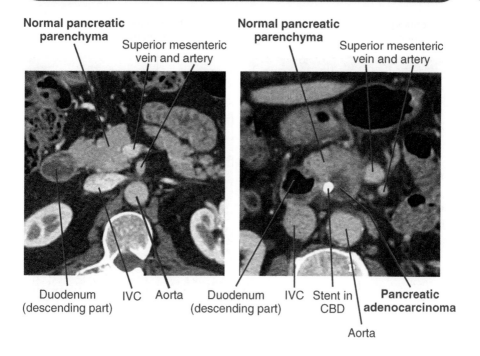

Contrast-enhanced axial CT images of the abdomen, normal on the left. The image on the right shows a patient who has a small pancreatic adenocarcinoma, visible as a hypoenhancing mass. Because of the intimate relationship between the pancreas and the common bile duct (CBD), patients with cancer in the head of the pancreas frequently present with obstructive jaundice. In this case the tumor had obstructed the CBD, which was treated by placing a stent in the duct to restore bile flow (the stent is radiopaque).

CT image of pancreatic cancer provided by Chandana Lall, MD, Indiana University School of Medicine, Indianapolis, IN.

GAS: Posthepatic jaundice (p. 326)
COA: Pancreatic cancer (p. 283)

Malrotation of the Small Bowel

Descending
duodenum

Descending
duodenum

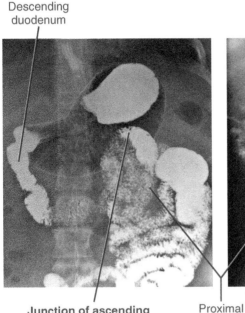
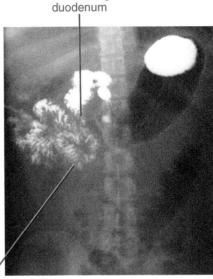

Junction of ascending
duodenum with jejunum,
indicating location of the
suspensory ligament of the
duodenum (ligament of Treitz)

Proximal
jejunum

In malrotation, there is no
ascending portion of the
duodenum or ligament of Treitz

Radiographs from barium upper GI examinations. In the right image, the jejunum is abnormally located on the patient's right side. Malrotation may present as an emergency in pediatric surgery with small-bowel obstruction and even catastrophic volvulus with strangulation of the small bowel. In those who have been asymptomatic with malrotation, diagnostic confusion (such as the patient exhibiting symptoms of appendicitis on the left side) or surgical errors may occur. In malrotation, there is no ascending portion of the duodenum or ligament of Treitz.

GAS: Malrotation and midgut volvulus (p. 313)
COA: Brief review of the embryological rotation of the midgut (pp. 258-259)

Obstructed Common Bile Duct

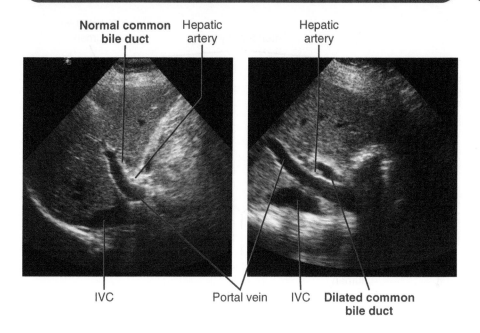

Normal common bile duct | Hepatic artery | Hepatic artery

IVC | Portal vein | IVC | **Dilated common bile duct**

Parasagittal ultrasound views of a barely visible normal-sized *(left)* and dilated *(right)* common bile duct. Dilatation of the duct implies distal obstruction, commonly caused by an impacted gallstone in the distal duct at the sphincter of Oddi, a tumor of the hepatopancreatic ampulla, a benign stricture of the distal duct, or a pancreatic cancer.

GAS: Gallstones (p. 326)
COA: Gallstones (pp. 286-287)

Gallstones

Anterior body wall

Gallbladder lumen

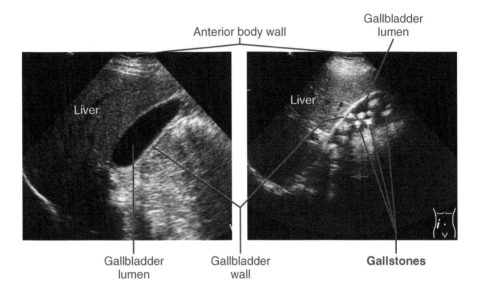

Liver

Liver

Gallbladder lumen

Gallbladder wall

Gallstones

Ultrasound images, normal on left; images are oriented along the long axis of the gallbladder. The gallstones reflect sound (bright echoes), causing a posterior acoustic shadow (indicated by red arrowheads on right image). To detect or exclude the presence of cholelithiasis, ultrasonography is the procedure of choice.

GAS: Gallstones (p. 326)
COA: Gallstones (pp. 286-287)

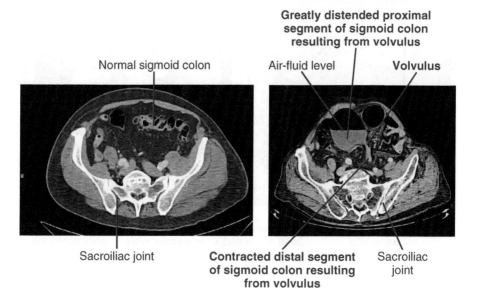

Normal sigmoid colon

Greatly distended proximal
segment of sigmoid colon
resulting from volvulus

Air-fluid level

Volvulus

Sacroiliac joint

**Contracted distal segment
of sigmoid colon resulting
from volvulus**

Sacroiliac
joint

Axial CT images at the level of the sacroiliac joints, with normal anatomy shown in the left image. The image on the right depicts a patient with a volvulus (twisting) in the sigmoid colon. The volvulus of the very mobile sigmoid colon results in distention of the colon proximal to the volvulus because its contents are prevented from entering the more distal colon segments. The air-fluid level is present because the patient is supine; the accumulated colonic fluids pool posteriorly.

GAS: Congenital disorders of the gastrointestinal tract (p. 313)
COA: Volvulus of the sigmoid colon (p. 261)

Appendicitis

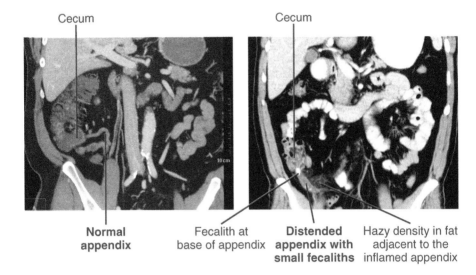

Cecum

Cecum

Normal appendix

Fecalith at base of appendix

Distended appendix with small fecaliths

Hazy density in fat adjacent to the inflamed appendix

Coronal CT images of the abdomen. The left image shows a patient with normal anatomy; here, the appendix is a thin structure surrounded by a normal fat. Appendicitis is associated with obstruction of the lumen, sometimes by a calcified fecalith *(right)*. An inflamed appendix is distended with inflammatory fluid, has a thickened wall, and is surrounded by edematous periappendiceal fat. CT has become the imaging procedure of choice when appendicitis is suspected in the adult patient. In pediatrics, ultrasonography is often used for diagnosis.

GAS: Appendicitis (p. 310)
COA: Appendicitis (pp. 259-260)

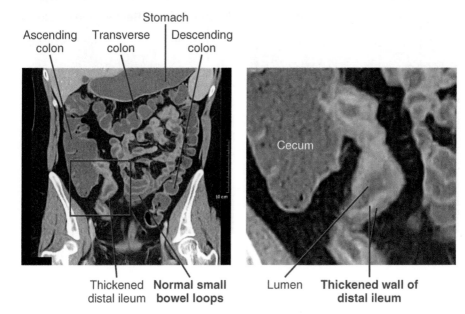

Stomach

Ascending colon   Transverse colon   Descending colon

Cecum

Thickened distal ileum   **Normal small bowel loops**   Lumen   **Thickened wall of distal ileum**

Coronal CT images of the abdomen showing hyperenhancement (indicating hyperemia) and thickening (mural edema and inflammatory hyperplastic changes) of the distal small bowel in this patient with acute Crohn's disease. The image on the right is a magnified view of the area shown by the red box in the image on the left. Note the very thin walls of normal small-bowel loops and of the colon. It is important to examine distended bowel loops for mural thickening because the wall thickness of contracted normal bowel segments may appear to be thickened. In the magnified view of the abnormal distal ileum shown on the right, mural thickening can clearly be seen to persist in this moderately distended segment.

GAS: Computed tomography (CT) scanning and magnetic resonance imaging (MRI) (p. 306)
COA: Colitis, colectomy, ileostomy, and colostomy (p. 260)

Ulcerative Colitis

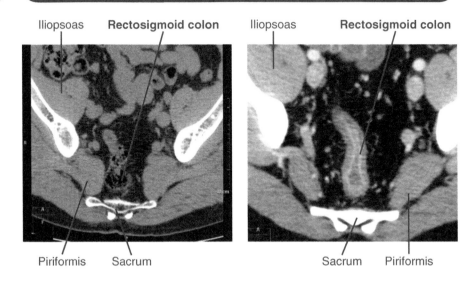

Iliopsoas | Rectosigmoid colon | Iliopsoas | Rectosigmoid colon

Piriformis | Sacrum | Sacrum | Piriformis

Slightly oblique axial CT images, normal on left. The image on the right depicts a patient with inflamed mucosa *(bright white lines)* and a thickened, edematous wall of the distal colon, the typical appearance of ulcerative colitis. Whereas ulcerative colitis most commonly begins in the distal colon and progresses proximally in a continuous distribution, Crohn's disease (see p. 91) is typically found in the distal ileum immediately adjacent to the ileocecal valve, and may involve noncontiguous segments of the gastrointestinal tract.

COA: Colitis, colectomy, ileostomy, and colostomy (p. 260)

Normal right ureter   Uterus   **Left ureter dilated proximal to the obstructing calculus**   Spleen

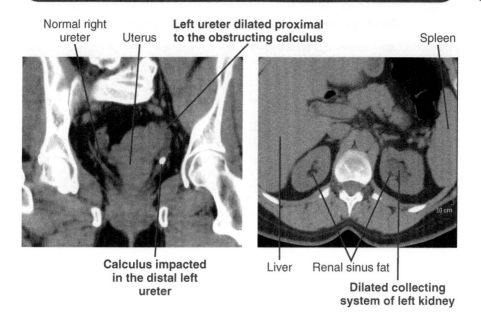

Calculus impacted in the distal left ureter   Liver   Renal sinus fat

Dilated collecting system of left kidney

For a suspected ureteral calculus, unenhanced CT imaging is the procedure of choice. The coronal CT image *(left)* shows a calculus impacted in the distal left ureter in a patient with acute left flank pain. The ureter proximal to the calculus and the intrarenal collecting system (axial CT image, *right*) are dilated.

GAS: Urinary tract stones (p. 361)
COA: Renal and ureteric calculi (p. 300)

Benign Prostatic Hypertrophy (BPH)

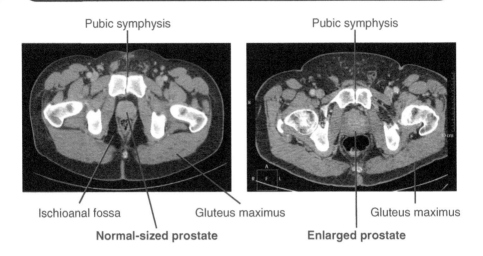

Pubic symphysis   Pubic symphysis

Ischioanal fossa   Gluteus maximus   Gluteus maximus

Normal-sized prostate   **Enlarged prostate**

Axial CT images of a patient with a normal *(left)* and an enlarged prostate gland *(right)*. BPH is commonly diagnosed via a digital rectal examination, but the enlargement is also readily visible on CT scans.

GAS: Prostate problems (pp. 451-452)
COA: Hypertrophy of prostate (p. 381)

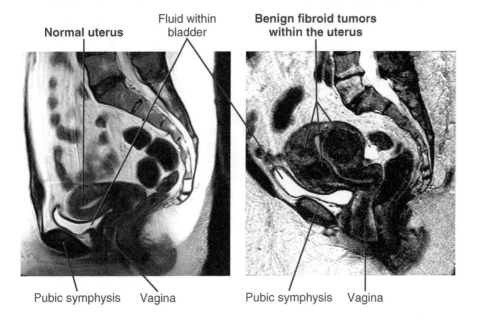

Normal uterus · Fluid within bladder · Benign fibroid tumors within the uterus

Pubic symphysis · Vagina · Pubic symphysis · Vagina

Sagittal MR images of the pelvis, normal on left. Fibroid tumors (leiomyomas) are apparent within the uterus in the image on the right. Such fibroid masses are very common in older women, have a strong familial predilection, and may enlarge rapidly during pregnancy as a result of hormonal influences. They rarely become malignant. Fibroids are most commonly diagnosed by ultrasonography; however, MR images provide a more definitive demonstration of uterine disease.

Bicornuate Uterus

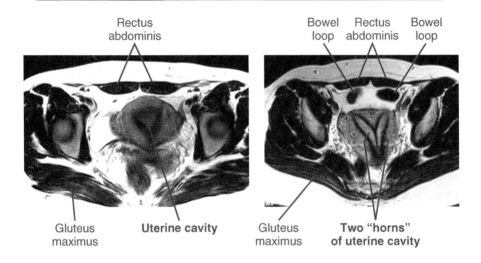

Rectus abdominis

Bowel loop   Rectus abdominis   Bowel loop

Gluteus maximus

Uterine cavity

Gluteus maximus

Two "horns" of uterine cavity

Axial MR images of a woman with a bicornuate uterus *(right)* compared with a woman with a normal uterus *(left)*. Müllerian duct anomalies are found in a wide spectrum of uterine developmental abnormalities with varied degrees of deformity, from an arcuate uterus to uterus didelphys. Approximately 10% of these are bicornuate. Although women with this uterine abnormality usually have little difficulty with conception, they have higher than normal rates of spontaneous abortion and preterm delivery.

COA: Bicornuate uterus (p. 392)

Transvaginal transducer Uterus Transvaginal transducer

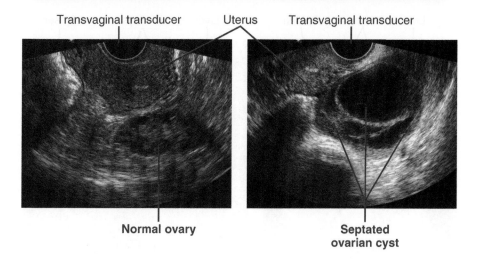

Normal ovary

Septated
ovarian cyst

Transvaginal ultrasound images of a patient with a septated ovarian cyst *(right)* and a patient with a normal ovary *(left)*. The anechoic (dark) spaces on the right represent fluid, and the bright curvilinear structures are septations within the cyst. The fluid in the cyst transmits sound well, resulting in bright echoes deep to it.

GAS: Imaging the ovary (p. 455)
COA: Laparoscopic examination of pelvic viscera (p. 397)

Ovarian Dermoid Cyst (Teratoma)

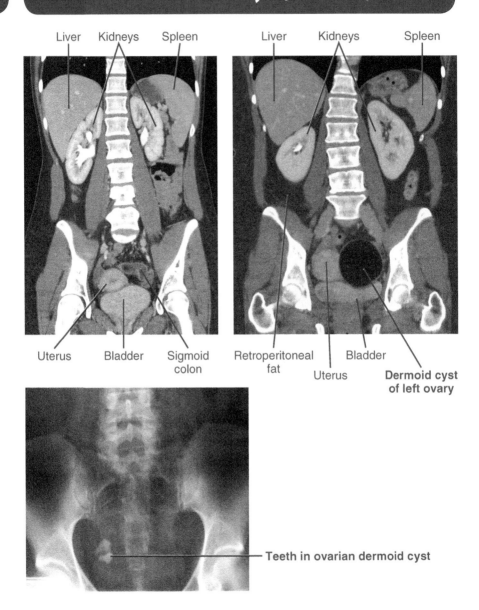

Coronal CT images *(top)* and AP radiograph *(bottom)*. The very low CT density of the pelvic mass in the upper right image (even darker than retroperitoneal fat) is consistent with lipid material; pathology revealed the dermoid cyst to contain oily liquid, hair follicles, and skin. These teratomas may also contain teeth *(bottom, different patient)*. Normal ovaries may be small and not apparent in CT images.

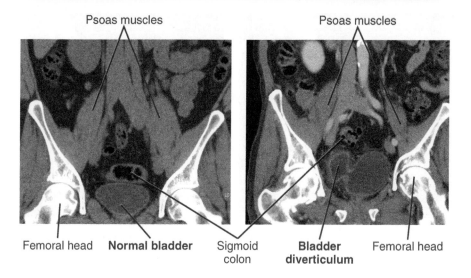

Psoas muscles — Psoas muscles

Femoral head · **Normal bladder** · Sigmoid colon · **Bladder diverticulum** · Femoral head

Coronal CT images of a patient with a bladder diverticulum *(right)* and a patient with a normal bladder *(left)*. A bladder diverticulum may be developmental and present in infancy, or may become apparent (as in this case) later in life. Then they are usually associated with bladder outlet obstruction such as prostatic hypertrophy. Although the lumen of the diverticulum is in communication with the rest of the bladder, emptying is incomplete, and stasis of urine in the diverticulum predisposes the patient to urinary tract infection.

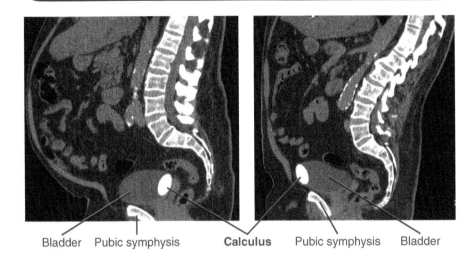

Bladder   Pubic symphysis   **Calculus**   Pubic symphysis   Bladder

Supine *(left)* and prone *(right)* sagittal CT images of a patient with a large bladder calculus. The calculus moves within the bladder as shown by its different locations when the patient changes position. Bladder calculi, more often than renal or ureteral calculi, are composed of uric acid. These may not be radiographically dense and therefore may not be apparent on radiographs. Unenhanced CT scanning, as shown here, is the procedure of choice for imaging bladder calculi.

GAS: Bladder stones (p. 443)

Testis with
normal vessels   Scrotal wall

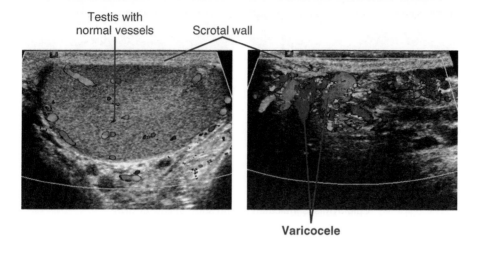

Varicocele

Ultrasound/Doppler examination of the scrotum. In the image on the left, normal vessels are depicted in and around the testis. In the study on the right, color Doppler imaging confirmed the clinical suspicion that a palpable mass was extratesticular and was a varicocele (collection of dilated intrascrotal veins of the pampiniform plexus).

COA: Varicocele (p. 215)

Epididymitis

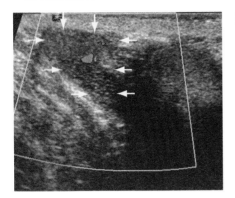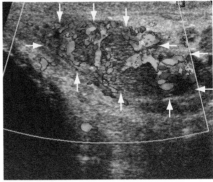

Epididymis is indicated by the short yellow arrows.

Ultrasound examination of the scrotum. Color flow Doppler sonography is ideal for demonstrating the increased blood flow that characterizes epididymitis, usually a unilateral condition. The hyperemia on the right side is obvious when compared with the normal epididymis on the left. Although testicular torsion was considered in the differential diagnosis in this patient, the ultrasound examination study indicated the presence of epididymitis as an explanation for the hemiscrotal pain.

Head of epididymis

Epididymal cyst

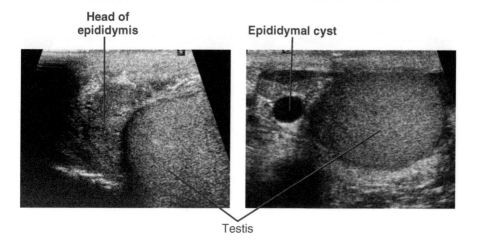

Testis

Approximately coronal views from ultrasound examination of the scrotum. When such an examination documents that an intrascrotal mass is within the testis, a malignant tumor is likely (see Testicular Tumor, p. 105). However, intrascrotal masses external to the testis are rarely malignant. Epididymal cysts in the head of the epididymis, as the one shown here, are indistinguishable from a spermatocele in ultrasound images.

GAS: Testicular tumors (p. 448)
COA: Spermatocele and epididymal cysts (p. 215)

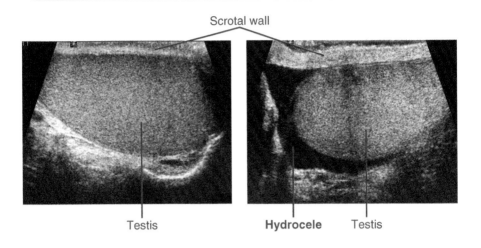

Scrotal wall

Testis

Hydrocele   Testis

Ultrasound examination of the scrotum showing a hydrocele in the image on the right. The serous fluid in the hydrocele is typically within the tunica vaginalis and may be associated with an indirect inguinal hernia. Ultrasound imaging has largely replaced transillumination for diagnosis of a hydrocele.

COA: Hydrocele of spermatic cord and/or testis (p. 212)

Scrotal wall

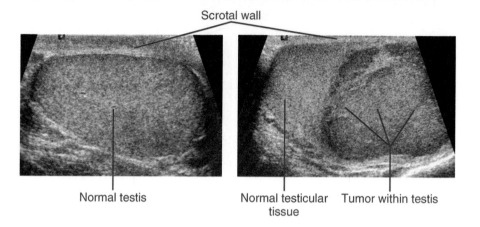

Normal testis

Normal testicular tissue

Tumor within testis

Ultrasound examination of the scrotum; in the testis shown on the right, ultrasonography confirmed that a clinically palpable mass was within the testis. After orchiectomy, pathology revealed a localized seminoma that had not metastasized to lymph nodes. Testicular tumors are the most common malignancy in males 15 to 35 years of age.

GAS: Testicular tumors (p. 448)
COA: Cancer of testis and scrotum (p. 215)

Testicular Torsion

Scrotal wall

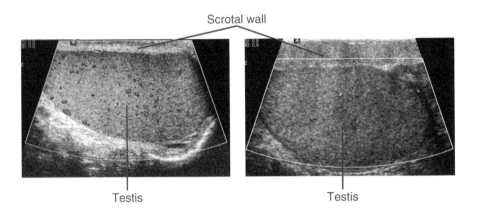

Testis   Testis

Duplex sonography (ultrasound with Doppler signal) of the scrotum (normal on left). The image on the right shows reduced color (Doppler signal), and therefore decreased blood flow compared to the normal control. Decreased or absent testicular blood flow indicates testicular torsion. This patient presented with hemiscrotal pain and this duplex sonogram enabled differential diagnosis from another common cause of hemiscrotal pain, epididymitis (p. 100).

COA: Torsion of spermatic cord (p. 214)

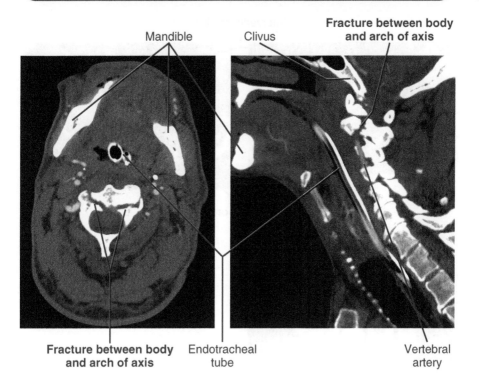

Mandible · Clivus · **Fracture between body and arch of axis**

Fracture between body and arch of axis · Endotracheal tube · Vertebral artery

Axial *(left)* and sagittal *(right)* CT images of the upper cervical vertebral column showing a traumatic bilateral fracture between the body of the axis and its arch. This fracture pattern is given the grim moniker of "hangman's fracture" because it is common in this method of execution. The dens of the axis is not visible because the plane of the right image is oblique, passing lateral to the dens.

GAS: Vertebral fractures (pp. 84-85)
COA: Fracture and dislocation of axis (pp. 459-460)

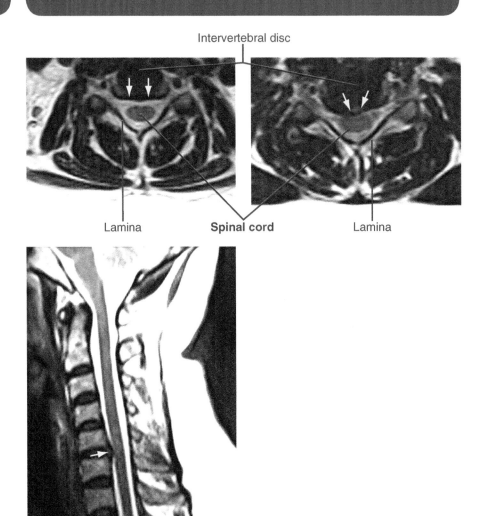

Intervertebral disc

Lamina — Spinal cord — Lamina

Axial MR images *(above)* and sagittal MR image *(below)* of the cervical spine. Compare the posterior disc margin in the individual with normal anatomy *(above left, yellow arrows)* to that in the patient with a herniated C5-C6 intervertebral disc *(above right and bottom images, yellow arrows)*. The herniated disc fragment protrudes into the spinal canal and impinges on the spinal cord. When a neurologic examination suggests significant neural impingement, MR imaging is the procedure of choice.

GAS: Herniation of intervertebral discs (p. 81)
COA: Herniation of nucleus pulposus (p. 476)

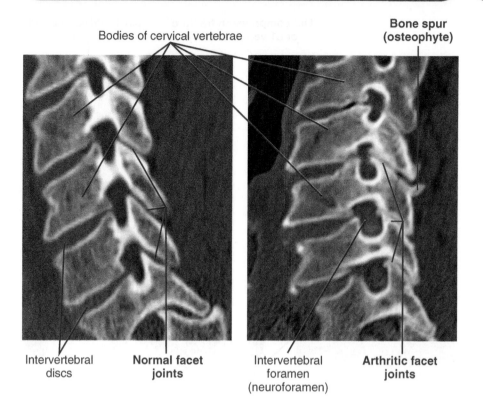

Oblique parasagittal CT images of a normal cervical spine *(left)* and of a patient with osteoarthritis of the cervical facet (zygapophyseal) joints *(right)*. In the right image, note the loss of articular cartilage (absence of joint space) between the superior and inferior articular processes (facets) and the presence of bone spurs (osteophytes). MR imaging is preferred for suspected disc disease, but CT is better for bone detail.

GAS: Joints (p. 82)
COA: Injury and disease of zygapophyseal joints (p. 480)

Vertebral Body Compression Fracture

Old compression fracture of L1 vertebral body | Spinal cord surrounded by CSF

L5 vertebra | Fluid in dural sac | **Subacute L3 compression fracture** | L5 vertebra | Fluid in dural sac

T1 (dark fluid, *left*) and T2 (bright fluid, *right*) sagittal MR images of the spine from T10 to the sacrum; both the L1 and L3 vertebrae are compressed when compared with the normal contours of the other vertebrae shown. However, the normal MR signal at L1 (same shade of gray as the normal vertebrae) indicates that this is an old fracture. The compression fracture of L3 shows edema (fluid signal) that is evidence of an acute or subacute bone injury.

GAS: Vertebral fractures (pp. 84-85)
COA: Fractures and dislocations of vertebrae (pp. 477-480)

BACK

Superior articular process

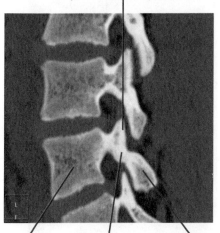

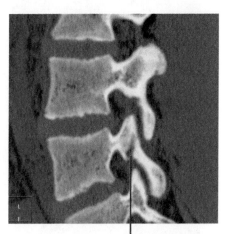

Vertebral body   **Intact pars interarticularis**   Inferior articular process   **Pars interarticularis fracture**

The high spatial resolution of these sagittal CT images provides the most accurate assessment of suspected pars interarticularis abnormalities, whether stress fracture, acute posttraumatic fracture, or old nonunited fracture (spondylolysis). Such pars defects may lead to the affected vertebra sliding anteriorly (spondylolisthesis; next page).

GAS: Pars interarticularis fractures (p. 85)
COA: Spondylolysis and spondylolisthesis (pp. 336-337)

Spondylolisthesis (Secondary to Pars Defect)

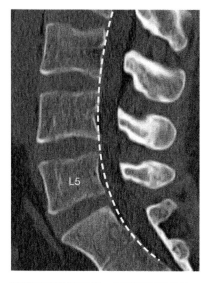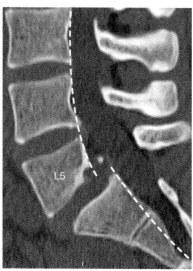

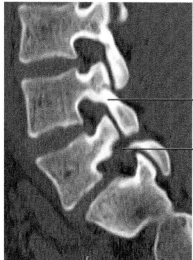

— Intact L4 pars interarticularis

— **L5 pars defect**

Normal midsagittal CT image *(top left)*, spondylolisthesis *(top right)*, and parasagittal CT image in the plane of the pars interarticularis in the same patient as shown in the upper right image *(bottom)*. Because of bilateral L5 pars interarticularis defects, the L5 vertebra slipped forward, out of normal alignment with the sacrum (compare alignment as indicated by the yellow dashes). The degree of slip, or listhesis, is often progressive.

GAS: Pars interarticularis fractures (p. 85)
COA: Spondylolysis and spondylolisthesis (pp. 336-337)

BACK

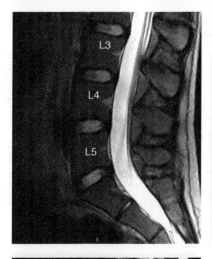

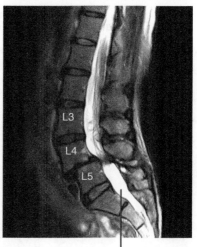

CSF in subarachnoid space

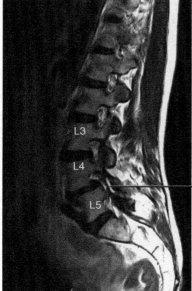

Joint effusion in arthritic facet joint

The upper left and upper right images are midline sagittal MR images, with the normal image on the left. The upper right image shows anterior displacement of the vertebral column superior to L5 (spondylolisthesis) because of L4-L5 facet joint degeneration (osteoarthrosis). In the parasagittal view on the bottom the bright fluid signal indicates the presence of a joint effusion in an arthritic facet joint. A degenerative listhesis is associated with spinal canal and neuroforaminal stenosis that may cause significant neural impingement.

GAS: Pars interarticularis fractures (p. 85)
COA: Fractures and dislocation of vertebrae (pp. 477-480)

Degenerative Spondylolisthesis (2)

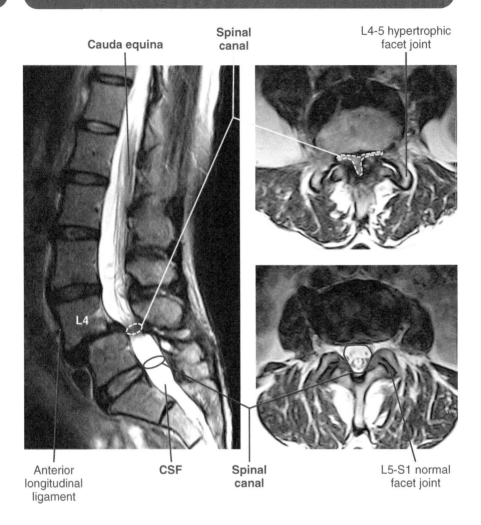

Cauda equina

Spinal canal

L4-5 hypertrophic facet joint

L4

Anterior longitudinal ligament

CSF

Spinal canal

L5-S1 normal facet joint

MR images; the sagittal view *(left)* shows anterior displacement of the vertebral column superior to L5 (spondylolisthesis) as a result of L4-L5 facet joint degeneration (osteoarthrosis), which is apparent in the upper right axial view compared with the normal facet joint at L5-S1 *(lower right)*.

GAS: Pars interarticularis fractures (p. 85)

COA: Fractures and dislocations of vertebrae (pp. 477-480)

COA: Injury and diseases of zygapophyseal joints (p. 480)

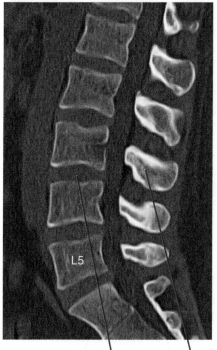 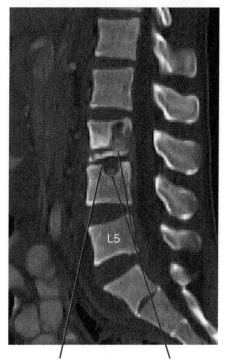

L3-4 intervertebral disc   Spinous process   **Collapsed L3-4 intervertebral disc**   **Erosions in vertebral bodies**

Sagittal CT images; infection in an intervertebral disc and adjacent lumbar vertebrae is shown on the right *(normal on the left)*. Evidence of the bone destruction associated with vertebral column osteomyelitis is very apparent in the CT image on the right. MR imaging would show changes in the bone marrow signal earlier than osseous changes shown in CT images. Vertebral column infections may spread to the adjacent psoas muscles as shown on p. 74.

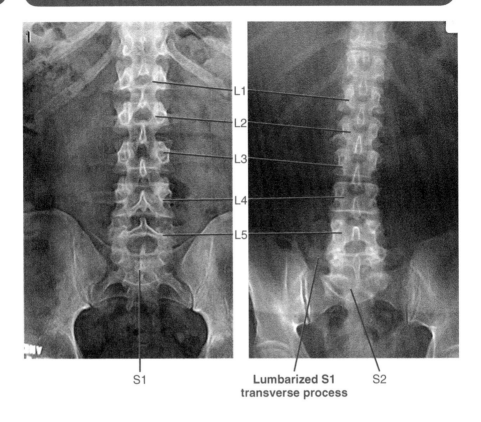

S1

Lumbarized S1
transverse process

S2

Congenital development of an increased number of lumbar vertebrae is typically an inconsequential finding in plain radiographs as shown here. Technically, however, if there are 12 true thoracic vertebrae (as there were in the case depicted on the right), then the apparent sixth lumbar vertebra is actually the first sacral vertebra (lumbarization of S1). This must be the case because the number of spinal nerves (and therefore the number of total vertebrae) is constant.

GAS: Variation in vertebral numbers (p. 78)
COA: Abnormal fusion of vertebrae (p. 462)

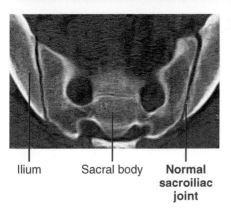 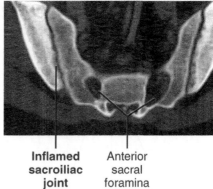

Ilium Sacral body **Normal sacroiliac joint** **Inflamed sacroiliac joint** Anterior sacral foramina

Slightly oblique axial CT images through the superior aspect of the sacrum. Sacroiliitis is an inflammation of one or both of the sacroiliac joints. Note the irregular articular surfaces of the joints in the pathologic image on the right compared with the healthy joints on the left. Sacroiliitis may be part of a more generalized spondyloarthropathy, such as psoriatic arthritis, reactive arthritis (Reiter syndrome), and ankylosing spondylitis, and may be associated with inflammatory bowel disease.

GAS: Common problems with the sacroiliac joints (p. 428)

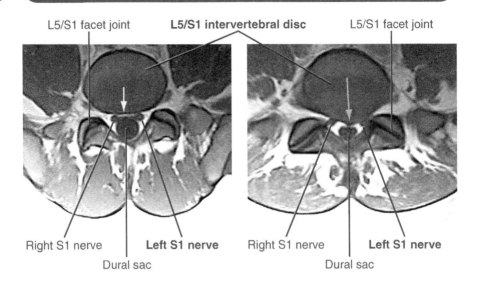

L5/S1 facet joint **L5/S1 intervertebral disc** L5/S1 facet joint

Right S1 nerve **Left S1 nerve** Right S1 nerve **Left S1 nerve**

Dural sac Dural sac

Axial MR images; patient with a herniated L5-S1 disc *(right)* and patient with a normal disc *(left)*. Note the posterior margin *(yellow arrow)* of the normal L5-S1 disc and the normal bright epidural fat surrounding the right and left S1 nerves. With disc herniation *(right)* the posterior disc protrusion *(orange arrow)* results in mild impingement on the right S1 nerve and compression of the left S1 nerve between the disc fragment and the facet joint. There is a loss ("effacement") of epidural fat around the nerve.

GAS: Herniation of intervertebral discs (p. 81)
COA: Herniation of nucleus pulposus (pp. 474-476)

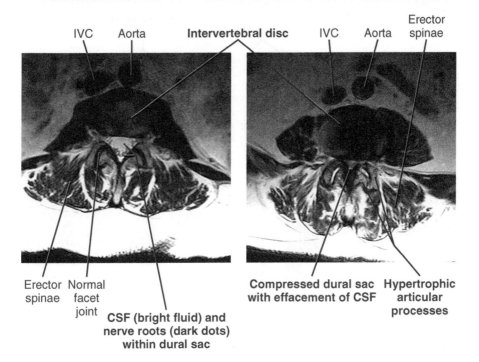

IVC   Aorta   **Intervertebral disc**   IVC   Aorta   Erector spinae

Erector   Normal
spinae   facet
          joint   **CSF (bright fluid) and
          nerve roots (dark dots)
          within dural sac**

**Compressed dural sac   Hypertrophic
with effacement of CSF   articular
                          processes**

Axial lumbar MR images; the image on the right shows compression of the nerves of the cauda equina attributable to hypertrophic (arthritic) facet joints and broad posterior bulging of a degenerated intervertebral disc. When the dural sac is reduced in diameter to a severe degree, nerve root ischemia may occur. For evaluation of suspected neural impingement, MR imaging is the procedure of choice in a patient who may be a surgical candidate.

GAS: Herniation of intervertebral discs (p. 81)
COA: Herniation of the nucleus pulposus (pp. 474-476)
COA: Injury and diseases of the zygapophyseal joints (p. 480)

Complete Transection of the Spinal Cord

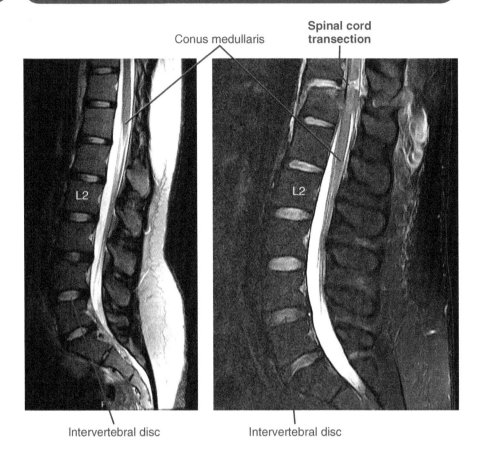

Sagittal MR images of the spine, normal *(left)* and a patient with a complete transection of the spinal cord at T11-T12 *(right)*. The T11 vertebral body is shifted slightly anterior to T12 because of complex unstable fractures of T12. For the cord to have been transected, there undoubtedly was a severe vertebral displacement at the time of injury that was reduced by the time this image was acquired. As a result of this severe injury, the patient became paraplegic and incontinent (fecal and urinary).

GAS: Vertebral fractures (pp. 84-85)
COA: Spinal cord injuries (p. 506)

BACK

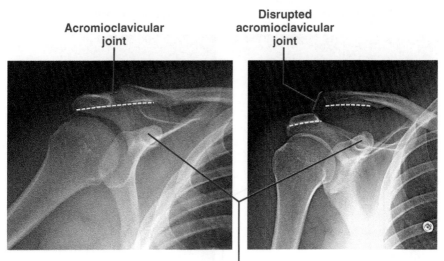

Acromioclavicular joint

Disrupted acromioclavicular joint

Coracoid process

AP radiographs of shoulder, normal on left. The right image shows a patient with an acromioclavicular (shoulder) joint separation. The integrity of this joint is dependent on the coracoclavicular ligaments, and these ligaments are typically torn in severe shoulder joint separations.

GAS: Fractures of the clavicle and dislocations of the acromioclavicular and sternoclavicular joints (pp. 673-674)

COA: Dislocation of acromioclavicular joint (pp. 813-814)

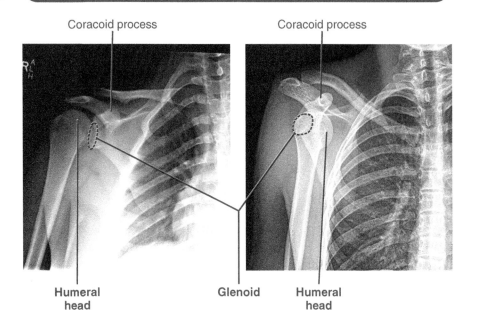

AP radiographs of a normal right shoulder *(left)* and one in which the humeral head is dislocated anteriorly *(right)*. In an ideal AP view of an anteriorly dislocated shoulder, the subcoracoid position of the humeral head confirms the anterior dislocation. However, a suboptimal AP view because of pain, body habitus, or other factors may result in some uncertainty about the position of the humeral head and a second projection is needed (see p. 123).

GAS: Dislocations of the glenohumeral joint (p. 674)
COA: Dislocation of glenohumeral joint (pp. 814-815)

Acromion process Coracoid process Acromion process Coracoid process

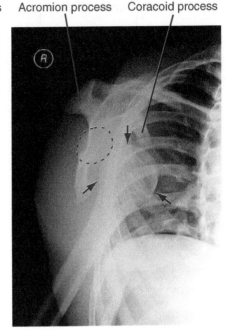

◄— Humeral head
⌐⌐⌐ Glenoid

"Y" lateral radiographic views of a normal shoulder *(left)* and a shoulder with an anterior dislocation of the humeral head *(right)*. Posterior dislocations are unusual, but do occur. With this projection, the direction of dislocation is confirmed. The vertical limb of the "Y" is the body of the scapula, and the upper two limbs are the acromion and coracoid processes.

GAS: Dislocations of the glenohumeral joint (p. 674)
COA: Dislocations of the glenohumeral joint (pp. 814-815)

Fractured Rim of Glenoid Fossa

Acromion   Coracoid process   Coracoid process

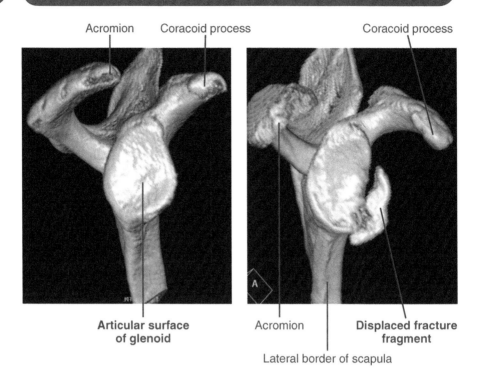

Articular surface   Acromion   **Displaced fracture**
of glenoid   **fragment**

Lateral border of scapula

CT reconstructions of the scapula from a lateral perspective; the humerus has been removed for clarity (normal on left). The patient with the glenoid fracture *(right)* had plain radiographs highly suspicious for a glenoid fracture, but the size, position, and even the number of fragments were not clear radiographically. It was also unclear how much of the articular surface was disrupted. CT scanning was requested for surgical planning. Similar 3D displays created from CT scanning are also very useful for fractures of the acetabulum, and other fractures resulting in anatomic derangement of complex osseous structures.

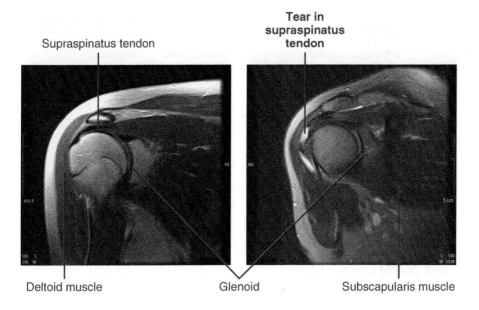

Supraspinatus tendon

Tear in supraspinatus tendon

Deltoid muscle · Glenoid · Subscapularis muscle

Tendons are dark and fluid is bright in these T2 MR images (normal on the left). The tear appears on the right image because fluid fills the gap between the torn parts of the tendon. Most rotator cuff tears involve the supraspinatus tendon, and most tears occur in tendons in which there has been some chronic degeneration.

GAS: Rotator cuff disorders (p. 675)
COA: Rotator cuff injuries (p. 712)

Superior Labrum, Anterior to Posterior (SLAP) Tear

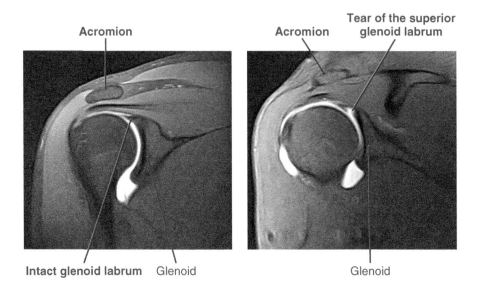

Acromion — Acromion — Tear of the superior glenoid labrum

Intact glenoid labrum — Glenoid — Glenoid

Coronal images from MR shoulder arthrography after intraarticular injection of gadolinium to increase contrast. The normal labrum is a dark triangular "cap" on the margin of the glenoid fossa. SLAP tears are very common in athletes who throw overhand (e.g., baseball pitchers and football quarterbacks). Although rotator cuff tears (see p. 125) are well visualized in routine shoulder MR images, shoulder MR arthrography is usually recommended for suspected labral tears.

COA: Glenoid labrum tears (p. 815)

Pectoralis major and minor

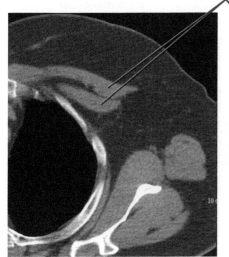 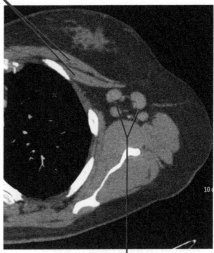

Enlarged axillary nodes

Axial CT images of patients without *(left)* and with *(right)* axillary adenopathy. Although enlarged axillary lymph nodes can often be palpated, this is sometimes difficult in larger patients. The presence or absence of enlarged axillary nodes can be confirmed with cross-sectional imaging, such as the CT scans shown here. CT imaging of the neck, chest, abdomen, and pelvis is also usually done during the staging of newly diagnosed lymphoma.

GAS: Breast cancer (p. 711)
COA: Carcinoma of the breast (pp. 104-105)

Dislocated Biceps Brachii Tendon

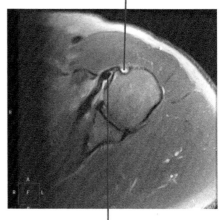

Lesser tubercle

Synovial fluid
in intertubercular
(bicipital) groove

Tendon of long head of the
biceps brachii, surrounded by
synovial fluid

Dislocated
biceps brachii
tendon

Axial MR images near the head of the humerus, normal on left. The tendon of the long head of the biceps brachii passes through the rotator interval between the insertions of the subscapularis and supraspinatus muscles to enter the intertubercular (bicipital) groove. Rotator cuff tears adjacent to the rotator interval may interrupt the "biceps sling mechanism," resulting in the tendon dislocation shown in the right image.

COA: Dislocation of tendon of the long head of biceps brachii (p. 741)

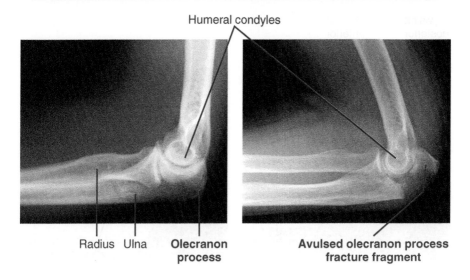

Humeral condyles

Radius   Ulna   **Olecranon process**

Avulsed olecranon process fracture fragment

Lateral elbow radiographs, normal on left. The image on the right shows a fracture of the proximal ulna with a separation of the olecranon process from the rest of the bone. Such an injury can result from a hard fall on the elbow and can cause injury to the ulnar nerve as it winds posteriorly around the medial epicondyle. Such widely distracted fracture fragments require internal fixation.

GAS: Elbow joint injury (p. 727)

Fracture of the Radial Head

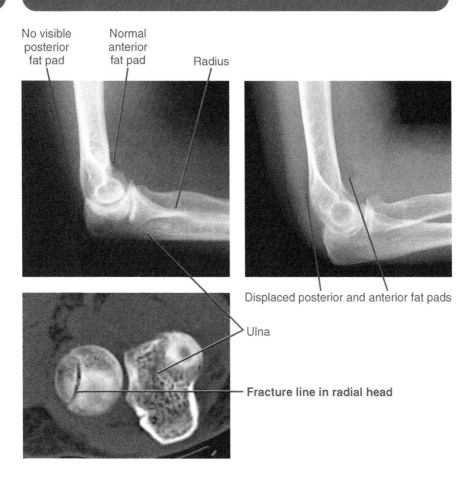

No visible posterior fat pad

Normal anterior fat pad

Radius

Displaced posterior and anterior fat pads

Ulna

Fracture line in radial head

On a lateral radiograph of a normal elbow *(top left)*, the posterior fat pad is not visible. With an elbow joint effusion or hemarthrosis, the posterior fat pad is displaced from the olecranon fossa and becomes visible. The anterior fat pad is displaced upwards *(top right)*. These radiographic signs are very important, because fine, nondisplaced fractures of the radial head may not be visible on initial radiographs. The patient shown here had a CT scan several weeks after injury (axial CT, *bottom*) that revealed the fracture line involving the articular surface of the radial head.

GAS: Elbow joint injury (p. 727)
GAS: Fracture of head of radius (p. 728)

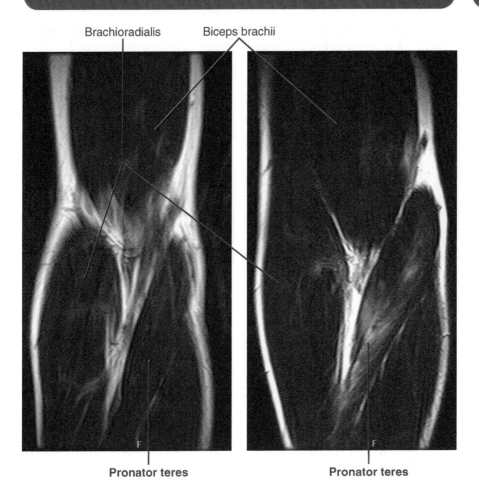

Brachioradialis   Biceps brachii

Pronator teres   Pronator teres

Coronal MR images, normal on the left. Note the grade 1 muscle strain of the pronator teres in the image on the right as shown by edema (bright fluid signal) in the muscle tissue, revealing its bipennate architecture. This image provides an explanation of the patient's elbow pain, especially upon flexion and pronation, while ruling out pathology that would need surgical repair.

COA: Muscle soreness and "pulled" muscles (p. 35)

Scaphoid Fracture

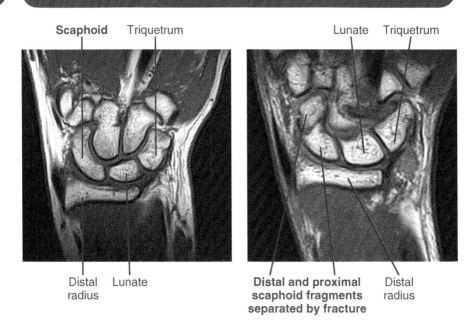

Coronal MR images of the wrist, normal on the left. The patient depicted in the right image has a fractured scaphoid, a common carpal bone fracture. When the scaphoid is fractured, pain is elicited upon palpation of the anatomic snuff box. Because the nutrient artery to the scaphoid usually enters the bone distally, the fragment that is proximal to such a transverse fracture may undergo avascular necrosis. In this case, the proximal fragment shows the same MR signal as normal bone, indicating that it is still viable, and the fracture may heal. Approximately 15% of nondisplaced scaphoid fractures are not visible on radiographs, but can be seen on an MR image.

GAS: Fracture of the scaphoid and avascular necrosis of the proximal scaphoid (p. 756)
COA: Fracture of scaphoid (p. 686)

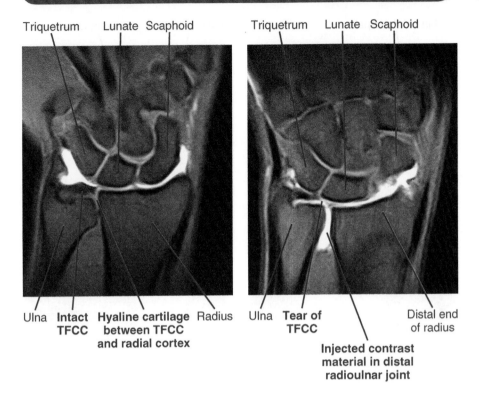

Triquetrum | Lunate | Scaphoid   Triquetrum | Lunate | Scaphoid

Ulna **Intact** **Hyaline cartilage** Radius   Ulna **Tear of** Distal end
     TFCC **between TFCC** **TFCC** of radius
          and radial cortex

**Injected contrast
material in distal
radioulnar joint**

Coronal MR images from wrist MR arthrography. In both cases, contrast material (bright fluid) has been injected into the radiocarpal joint. In the normal case *(left)* the TFCC, scapholunate, and lunotriquetral ligaments are intact, so the injected contrast material is confined to the radiocarpal joint. The apparent gray "gap" between the normal TFCC and the radius is filled by cartilage. On the right, the attachment of the TFCC to the radius is torn, allowing the injected contrast material to enter the distal radioulnar joint.

Colles Fracture

First metacarpal

First metacarpal   Head of ulna

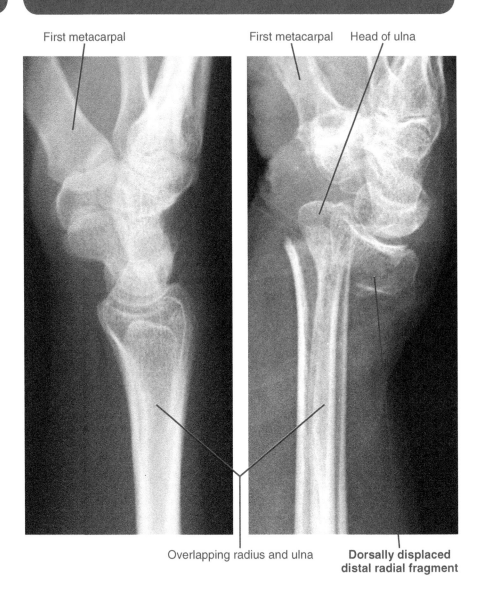

Overlapping radius and ulna

Dorsally displaced distal radial fragment

Lateral wrist radiographs, normal on left. In a typical Colles fracture (*right*), the distal radius is fractured transversely approximately 1 cm from the articular surface. On lateral radiographs, there is dorsal angulation and displacement of the distal radial fragment. Colles fractures typically occur in elderly osteoporotic patients who fall on their outstretched hands. Such falls in younger patients are more likely to result in a comminuted fracture that involves the articular surface of the radius.

GAS: Fractures of the radius and ulna (p. 734)
COA: Fractures of radius and ulna (pp. 685-686)

First metacarpal

First metacarpal   Head of ulna

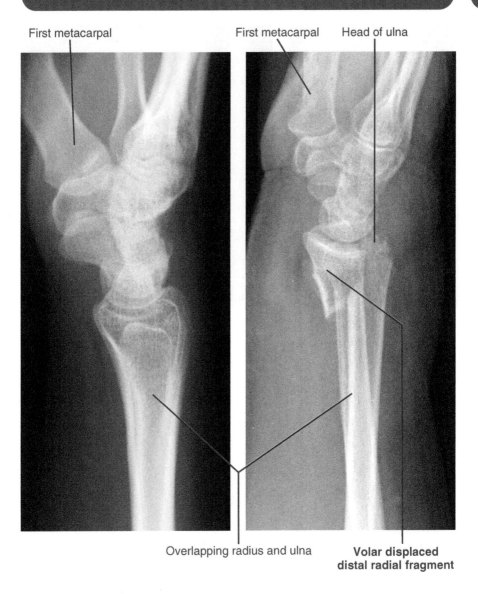

Overlapping radius and ulna   **Volar displaced
distal radial fragment**

Lateral wrist radiographs, normal on left. The Smith fracture *(right)* is often referred to as a reverse Colles fracture (see p. 134). It results from a fall on a flexed wrist. The distal radial fragment is displaced and/or angulated in a volar (ventral) direction. As with most similar injuries, treatment by immobilization, closed reduction, or open reduction and internal fixation depends upon the degree of comminution, angulation deformity, and fracture fragment displacement.

Boxer's Fracture

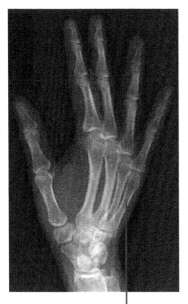

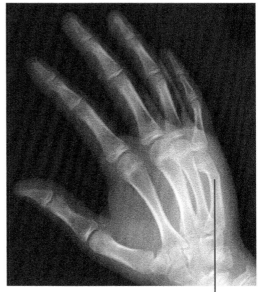

Normal fifth
metacarpal

Fractured fifth
metacarpal

Oblique radiographs of a normal hand *(left)* and one that has sustained an acute fracture of the fifth metacarpal ("boxer's fracture") *(right)*. As shown in this case, these fractures are commonly nondisplaced but moderately angulated.

COA: Fracture of metacarpals (p. 687)

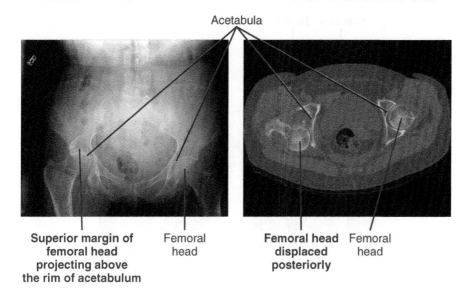

Acetabula

Superior margin of
femoral head
projecting above
the rim of acetabulum

Femoral
head

**Femoral head
displaced
posteriorly**

Femoral
head

AP radiograph *(left)* and axial CT image *(right)* of a patient with a right posterior hip dislocation, and fracture of the posterior rim of the acetabulum. In the radiograph, note the overlap of the right femoral head with the superior margin of the acetabulum. The axial CT image depicts more graphically the posterior displacement of the right femoral head. CT images superior and inferior to this level showed numerous fracture fragments of the posterior wall of the acetabulum.

GAS: Pelvic fractures (pp. 528-529)
COA: Dislocation of hip joint (pp. 660-661)

Metastatic Tumor of Acetabulum

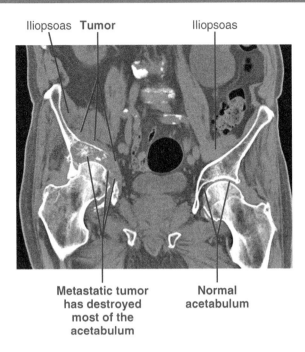

Iliopsoas   **Tumor**   Iliopsoas

Metastatic tumor   Normal
has destroyed   acetabulum
most of the
acetabulum

Coronal CT image showing destructive metastatic disease that has destroyed much of the acetabular fossa. This is a florid example of a "lytic" bone tumor that decreases the radiographic density of bone. Some tumors, such as metastatic breast and prostate carcinomas, may be "blastic" and cause increased bone density. CT is the procedure of choice to demonstrate the type of cortical destruction shown here, but MR imaging is best for showing tumor infiltration of bone marrow.

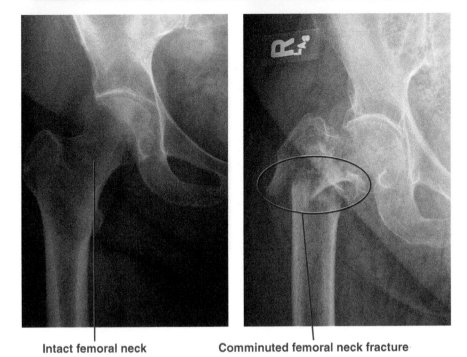

Intact femoral neck   Comminuted femoral neck fracture

AP hip radiographs (normal on left, fractured femur on right). "Hip fractures" are fractures of the proximal femur, usually the femoral neck. Other proximal femur fractures that are considered "hip fractures" include intertrochanteric femoral fractures and fractures of the greater trochanter. Isolated fractures of the lesser trochanter are highly suspicious for the presence of underlying tumor infiltration of the bone. In the case presented above, there are three major fracture fragments: 1) femoral shaft; 2) femoral head and most of the femoral neck; and, 3) greater trochanter. Femoral neck fractures are associated with osteoporosis, especially in elderly women.

GAS: Femoral neck fractures (p. 532)
COA: Fractures of femoral neck (pp. 659-660)

Degenerative Joint Disease, Hip

Normal articular
cartilage (joint space)

Eroded and thinned articular
cartilage (joint space)

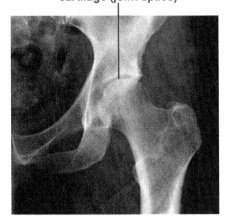 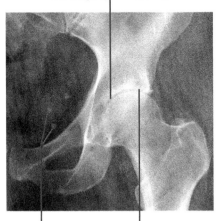

Phleboliths   Degenerative
subcortical cyst

Radiographs of a normal hip *(left)* and a patient with degenerative joint disease in the left hip joint *(right)*. Note the loss of articular cartilage and the presence of a subcortical degenerative cyst of the acetabulum in the arthritic hip. More advanced cases may show increased subchondral bone density in the acetabulum and femoral head, and marginal osteophytes. An incidental (and very common) finding is the presence of pelvic phleboliths, which are round calcifications at venous valves.

GAS: Degenerative joint disease/osteoarthritis (p. 582)
COA: Degenerative joint disease (pp. 28-29)

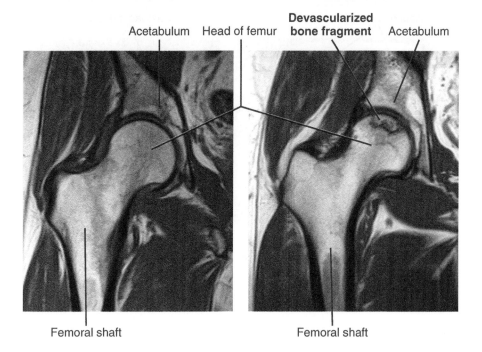

Acetabulum   Head of femur   **Devascularized bone fragment**   Acetabulum

Femoral shaft   Femoral shaft

Coronal MR images of the hip joint, normal on the left. On the right, the femoral head shows avascular necrosis (AVN) (note that the devascularized bone segment is demarcated by a subcortical dark line). In the adult, this is commonly a complication of steroid use. MR imaging is more sensitive than CT in detecting early-stage AVN, but later-stage disease that results in collapse of the articular surface is very well visualized on CT.

GAS: Femoral neck fractures (p. 532)
COA: Fractures of femoral neck (pp. 659-660)

Iliopsoas Bursitis

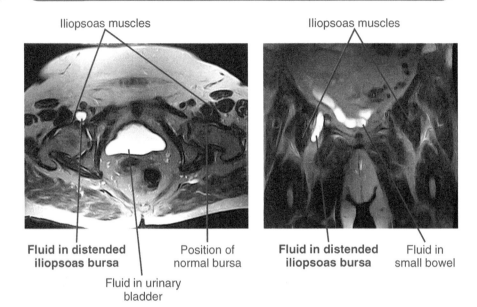

Iliopsoas muscles   Iliopsoas muscles

Fluid in distended iliopsoas bursa   Position of normal bursa   **Fluid in distended iliopsoas bursa**   Fluid in small bowel

Fluid in urinary bladder

Axial *(left)* and coronal *(right)* MR images of the same patient. A fluid-distended iliopsoas bursa is evident on the patient's right side compared to the normal condition on the left. Note that in this patient, distention of the bursa changes the apparent position of the bursa because it is no longer confined to a location between the iliopsoas tendon and the femoral neck.

Profunda
femoral arteries

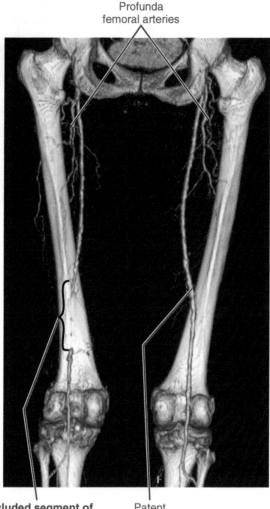

**Occluded segment of
femoral/popliteal artery**   Patent
femoral artery

Posterior view of lower limb CT arteriogram with 3D reconstruction. The left femoral artery is occluded by atherosclerotic plaque as it passes through the adductor hiatus to become the popliteal artery. Posterior to the popliteal fossa beyond the obstructed segment, the popliteal artery is filled by arterial collaterals such as the descending branches of the lateral circumflex femoral and the descending genicular arteries.

GAS: Peripheral vascular disease (p. 572)
COA: Popliteal pulse (p. 604)

Deep Venous Thrombosis

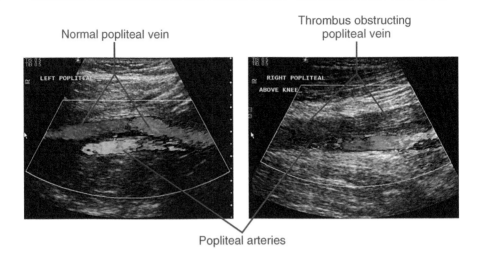

Normal popliteal vein

Thrombus obstructing popliteal vein

Popliteal arteries

Duplex sonography (ultrasound with Doppler signal) of the popliteal fossa in a normal right lower leg *(left)* compared with the image of a thrombosed popliteal vein *(right)*. The comparison reveals that no color Doppler signal could be demonstrated in the thrombosed vein. Posterior is at the top of both images.

GAS: Deep vein thrombosis (p. 544)

COA: Varicose veins, thrombosis, and thrombophlebitis (p. 540)

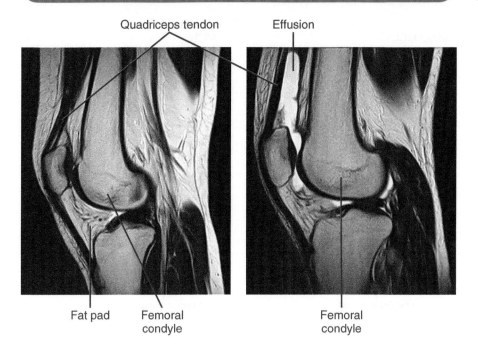

Quadriceps tendon   Effusion

Fat pad   Femoral   Femoral
           condyle   condyle

Sagittal MR images of a normal knee *(left)* and a knee with joint effusion (fluid within the joint) deep to the quadriceps tendon *(right)*. Such effusions usually accompany knee injury, or may indicate the presence of inflammatory arthritis.

GAS: Examination of the knee (p. 583)
COA: Aspiration of knee joint (p. 664)

Medial (Tibial) Collateral Ligament (MCL) Tear

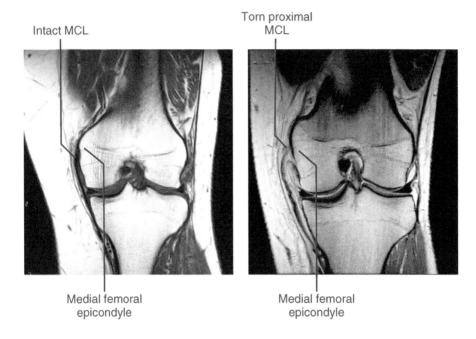

Intact MCL

Torn proximal MCL

Medial femoral epicondyle

Medial femoral epicondyle

Coronal MR images of the knee, normal on the left. The femoral attachment of the MCL is torn in the image on the right. Such tears cause periosteal stripping from the medial epicondyle, with subsequent calcification/ossification of the injured tissue. For approximately 100 years, the presence of an osseous spur projecting caudally from the medial epicondyle, known as Pellegrini-Stieda syndrome, has been recognized as evidence of an old tear of the MCL.

GAS: Soft tissue injuries of the knee (p. 582)
COA: Knee joint injuries (pp. 662-663)

Femoral
condyle

Anterior horn,
medial meniscus

**Posterior horn,
medial meniscus**

**Horizontal tear in posterior
horn, medial meniscus**

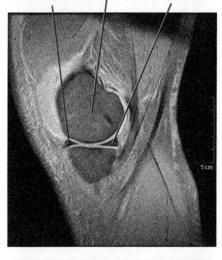

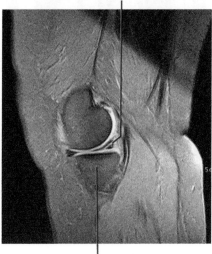

Tibial plateau

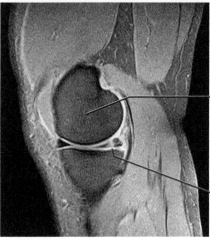

Femoral condyle

**Vertical tear in
posterior horn,
medial meniscus**

Sagittal MR images; the horizontal *(top right)* and vertical *(bottom)* tears in the posterior horn of the respective medial menisci appear as a bright disruption in the normally dark meniscus *(top left)*. Horizontal tears tend to be associated with degenerative changes in the joint whereas vertical tears tend to be associated with athletic injuries.

GAS: Soft tissue injuries to the knee (p. 582)
COA: Knee joint injuries (pp. 662-663)

Quadriceps Tendon Tear

Quadriceps
tendon

**Tear in the
quadriceps tendon**

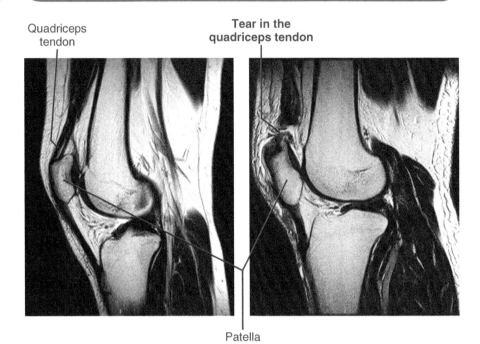

Patella

Sagittal MR images of the knee, normal on left. The right image shows a
patient with a complete tear of the quadriceps tendon. The quadriceps tendon attaches to the superior margin of the patella. A patient with a complete
quadriceps tendon tear cannot actively extend the knee. In this case the knee
was passively extended by the technologist who positioned the patient for
the MR scan.

GAS: Muscle injuries to the lower limb (p. 569)

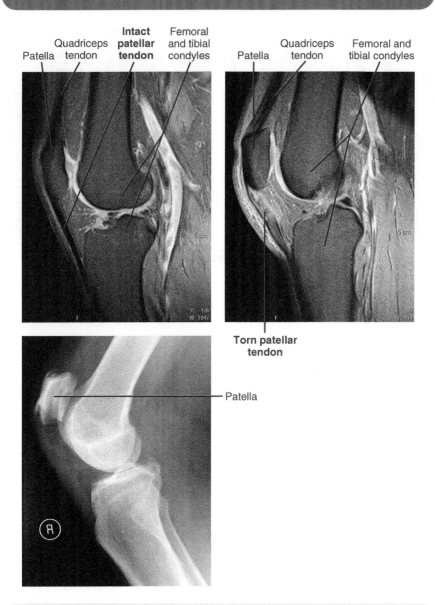

Sagittal MR images *(top)*; the upper right image shows a torn patellar tendon whereas the intact tendon is a continuous dark structure in the normal knee shown on the left. The lateral radiograph on the bottom shows an abnormally increased distance of the patella from the tibial tuberosity ("patella alta") in this case of patellar tendon tear. Patella alta may also occur with a developmentally long patellar tendon, and predisposes to patellar instability. "Patella infera" refers to an abnormally short patellar tendon.

Anterior Cruciate Ligament Tear

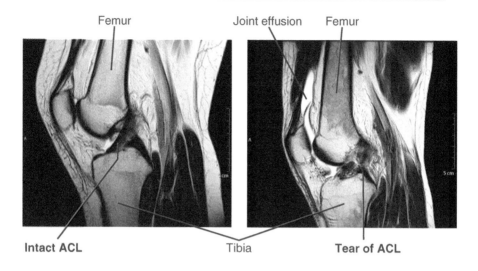

Femur   Joint effusion   Femur

Intact ACL   Tibia   **Tear of ACL**

Sagittal MR images; anterior is to the left. Visible discontinuity of the anterior cruciate ligament (ACL), as shown on the right, is definitive for diagnosis of a tear. Radiography and CT scanning provide more bone detail than MR images, especially of cortical surfaces, and are important for showing soft tissue calcifications. However, for clinically suspected internal derangements of the knee (such as cruciate ligament or meniscal tears), MR imaging is the procedure of choice.

GAS: Soft tissue injuries to the knee (p. 582)
COA: Knee joint injuries (pp. 662-663)

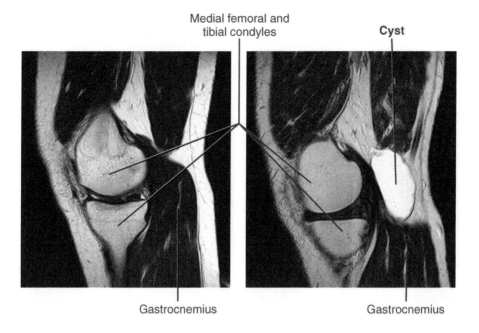

Medial femoral and
tibial condyles

Cyst

Gastrocnemius

Gastrocnemius

Sagittal MR images of the knee, normal on the left. The right image depicts a large popliteal (Baker) cyst, which is typically a synovial fluid–filled sac that is an expansion of the semimembranosus-gastrocnemius bursa. A large popliteal cyst may be painful and restrict range of motion. Most of these cysts are a result of increased joint fluid and are associated with joint disease. These cysts may be demonstrated on ultrasound or MR images.

COA: Popliteal cysts (p. 665)

Loss of articular cartilage (joint space) in medial compartment

Marginal osteophyte arising from medial femoral condyle

Normal joint spaces

Head of fibula   Tibia

Normal articular cartilage (joint space) in lateral compartment

Marginal osteophyte on tibial plateau

AP radiographs of a patient with degenerative joint disease in the medial knee compartment *(right)* compared with a normal knee *(left)*. Note the loss of articular cartilage and the increased subchondral bone density in the arthritic joint compartment as well as the marginal osteophytes at the medial aspect of the knee.

GAS: Degenerative joint disease/osteoarthritis (p. 582)
COA: Degenerative joint disease (pp. 28-29)

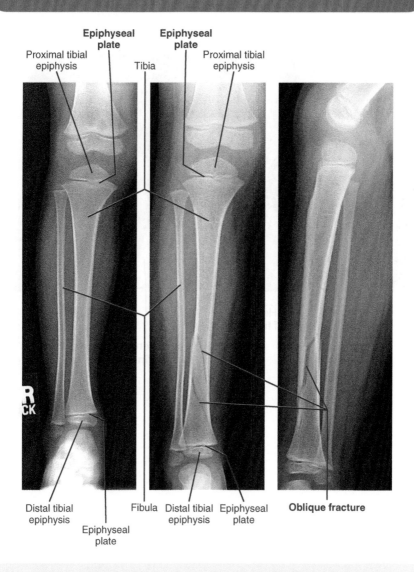

AP and lateral radiographs of a child with an acute oblique fracture of the tibial shaft *(middle and right)*, compared with AP radiograph of a child with a normal tibial shaft *(left)*. Note the epiphyseal (growth) plates in these immature bones that should not be misinterpreted as fractures. The proximal of the two fine linear radiolucencies on the AP view is the fracture through the posterior cortex, and the distal radiolucency is the fracture through the anterior cortex (see lateral view that shows the oblique fracture involving the posterior cortex more proximally than the anterior cortex).

COA: Bone growth and the assessment of bone age (p. 24)
COA: Tibial fractures (pp. 527-528)

Pes Anserinus Bursitis

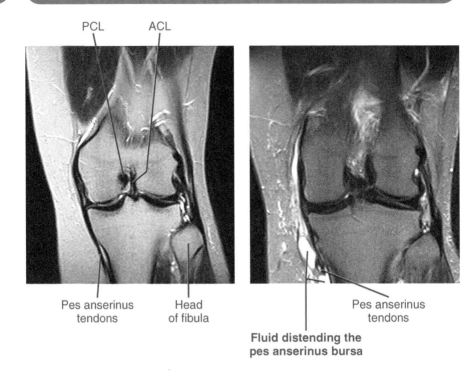

PCL   ACL

Pes anserinus   Head
tendons   of fibula

Pes anserinus
tendons

**Fluid distending the
pes anserinus bursa**

Coronal MR images of the knee, normal on left. Pes anserinus bursitis (bright fluid) is shown in the right image, which is inflammation of the pes anserinus bursa. This bursa is located between the tendons of the pes anserinus (sartorius, gracilis, and semitendinosus tendons) and the medial cortex of the tibia. When this bursa is distended, it often extends medial to the tendons, as shown here. Inflammation of this bursa is associated with pain, swelling, and point tenderness. Normal bursae have only trace amounts of fluid, not usually evident on imaging.

Tibia

Calcaneal
tendon

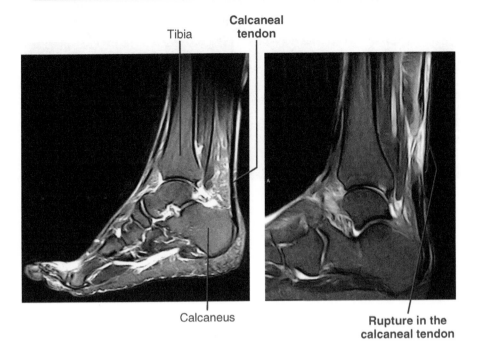

Calcaneus

Rupture in the
calcaneal tendon

T2 MR sagittal images of the ankle. The calcaneal tendon (frequently called "Achilles tendon," one of the more interesting eponyms in anatomy) is often ruptured as a sports injury when a person makes a sudden change in motion (e.g., abrupt stop).

COA: Ruptured calcaneal tendon (p. 607)

Calcaneal Fracture

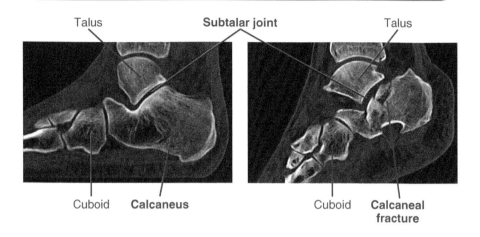

Talus   **Subtalar joint**   Talus

Cuboid   **Calcaneus**   Cuboid   **Calcaneal fracture**

Sagittal CT images of a patient with a comminuted calcaneal fracture *(right)* compared with that of a patient with normal anatomy *(left)*. The presence of multiple fracture lines, indicating a comminuted fracture, is common in these injuries. As in the case shown here, when the articular surface of the calcaneus at the subtalar joint is severely disrupted, a disabling subtalar joint arthritis may result if the deformity is not corrected.

COA: Calcaneal fractures (p. 529)

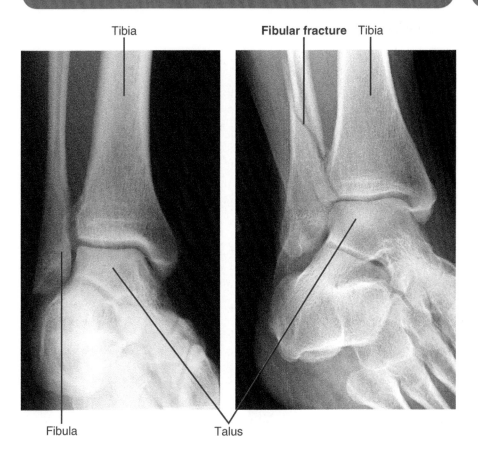

Tibia · Fibular fracture · Tibia · Fibula · Talus

Oblique radiographs ("mortise view") of the ankle, normal on left. The radiograph on the right shows a typical ankle fracture. Most ankle injuries result from inversion of the ankle and involve lateral structures, usually the anterior talofibular ligament or distal fibula. There are a large variety of ankle fractures, classified by the location and number of fractures. The stability of the ankle joint depends upon the location of the fracture in relation to the joint and the integrity of the syndesmoses at the ankle, the distal tibiofibular ligaments.

GAS: Ankle fractures (p. 608)
COA: Ankle injuries (pp. 665-666)

Medial malleolus

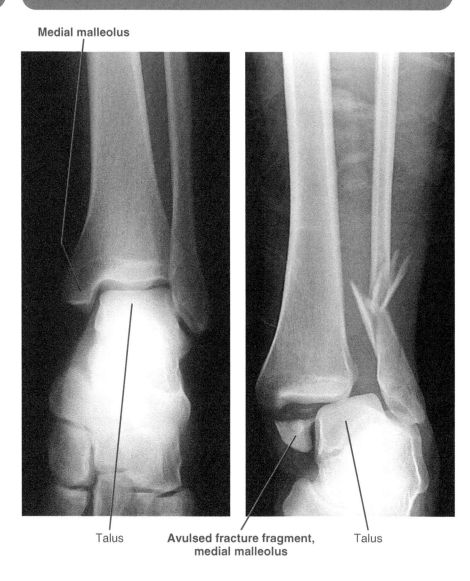

Talus

Avulsed fracture fragment, medial malleolus

Talus

AP radiographs of the ankle, normal on left. With eversion injuries, a fracture of the medial malleolus may occur, as shown here on the right. The fibula is commonly fractured, at a variety of anatomic levels, along with the medial malleolus. The stability of the ankle mortise depends upon the position and angle of the medial malleolar fracture and associated ligamentous and osseous injuries. In the case on the right, the fragment of the medial malleolus is displaced and the talus has been dislocated laterally. This injury has obviously disrupted the ankle mortise and will be treated with open reduction and internal fixation.

GAS: Ankle fractures (p. 608)
COA: Fibular fractures (p. 528)

Normal anterior talofibular ligament

Thickened and torn anterior talofibular ligament

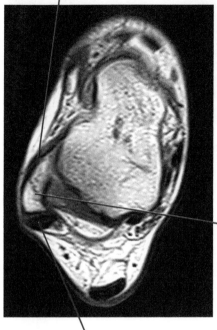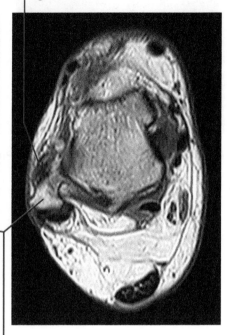

Tendons of fibularis (peroneus) longus and brevis

Lateral malleolus

Axial MR images of the ankle, normal on the left and sprained ankle on the right. Most ankle sprains are inversion injuries in which the first ligament to be injured is the anterior talofibular ligament, followed by the calcaneofibular ligament and posterior talofibular ligament. In a severe injury, the distal tibiofibular joint (which is syndesmotic) is ruptured, leading to widening of the ankle mortise and resulting in ankle instability.

GAS: Ankle fractures (p. 608)
COA: Ankle injuries (pp. 665-666)

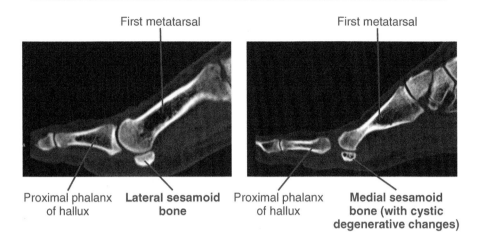

First metatarsal   First metatarsal

Proximal phalanx of hallux   **Lateral sesamoid bone**   Proximal phalanx of hallux   **Medial sesamoid bone (with cystic degenerative changes)**

CT scans along the long axis of the foot of the same patient. The lateral sesamoid bone *(left)* is normal whereas the medial sesamoid *(right)* has a cyst within it. This cystic condition, along with sesamoiditis and fractures, is a source of pain in the forefoot.

COA: Fracture of sesamoid bones (p. 531)

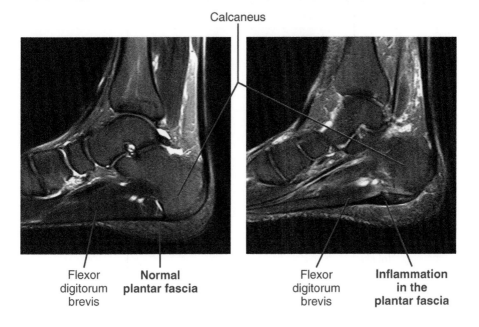

Calcaneus

| Flexor digitorum brevis | **Normal plantar fascia** | Flexor digitorum brevis | **Inflammation in the plantar fascia** |

MR sagittal images of the foot, normal on left. The right image shows inflammation of the plantar fascia. Note the thickened proximal plantar fascia near the calcaneus *(right)* compared with normal plantar fascia *(left)*. The normal fascia is very dark; pathology within the abnormal fascia is shown as a brighter shade in these images. Plantar fasciitis commonly causes pain along the inferomedial aspect of the foot, especially near the calcaneus, upon arising in the morning.

COA: Plantar fasciitis (p. 624)

Radionuclide Bone Scan

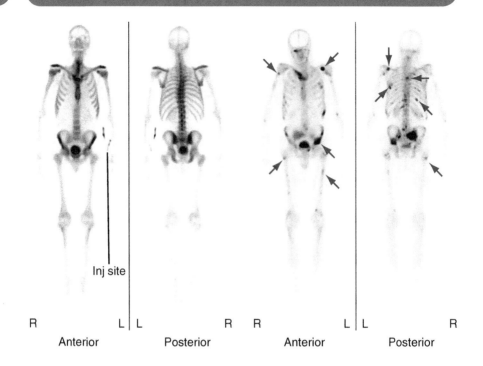

Inj site

R   L | L   R   R   L | L   R

Anterior   Posterior   Anterior   Posterior

**Arrows highlight several of many
foci of abnormal skeletal uptake (activity)**

Nuclear medicine scans generally provide information about physiologic, not anatomic, changes in disease. The left pair of images shows a normal skeletal distribution of the radiopharmaceutical that had been injected intravenously 3 hours before this scan (note that some adhered to the wall of the vein that had been used for injection). The images on the right show many skeletal foci of abnormally increased uptake. Scans like these are good examples of studies that may be sensitive (in detecting disease), but not specific. Anything causing increased bone mineral turnover, such as healing fracture, tumor, or infection, could cause such increased uptake. However, the pattern of activity shown here, in a patient previously diagnosed with a cancer that often spreads to bone, is almost certain to indicate widespread skeletal metastases.

GAS: Nuclear medicine imaging (pp. 10-11)
COA: Nuclear medicine imaging (p. 70)

Index

Note: Page numbers followed by *f* refer to figures

A

Abdominal adenopathy, 72, 72*f*
Abdominal aortic aneurysm, 73, 73*f*
Abscess, psoas, 74, 74*f*
Acetabulum, 137*f*, 141*f*
    fracture of, hip dislocation with, 137, 137*f*
    metastatic tumor of, 138, 138*f*
    normal, 138*f*
Achilles tendon tear, 155, 155*f*
Acoustic meatus, external, 24*f*
Acromioclavicular joint, separation of, 121, 121*f*
Acromion, 124*f*
Adenocarcinoma, pancreatic, 85, 85*f*
Adenoma, benign pleomorphic, of parotid gland, 25
Adenopathy, abdominal, 72, 72*f*
Adult polycystic kidney disease, 84, 84*f*
Air
    in esophagus, 47*f*
    in external auditory canal, 28*f*
Anastomosis
    of caput medusae with great saphenous vein,
        70, 70*f*
    of internal mammary artery to coronary artery,
        55, 55*f*
Aneurysm
    aortic
        abdominal, 73, 73*f*
        ascending, 44, 44*f*
    of internal carotid artery, 14, 14*f*, 15, 15*f*
Angiography
    of coronary angioplasty, 51, 51*f*
    of internal carotid artery aneurysm, 14, 14*f*
Angioplasty, coronary, 51, 51*f*
Ankle fracture, 157, 157*f*
Ankle sprain, 159, 159*f*
Anterior cruciate ligament
    intact, 150*f*
    tear in, 150, 150*f*
Anterior horn, medial meniscus, 147*f*
Aorta
    ascending, 42*f*, 47*f*
    descending, 36*f*, 42*f*, 47*f*
Aortic aneurysm
    abdominal, 73, 73*f*
    ascending, 44, 44*f*
Aortic arch, 43*f*, 44*f*, 45*f*, 48*f*, 66*f*
    right, 46, 46*f*
Aortic valve stenosis, 52, 52*f*
Appendicitis, 90, 90*f*
Appendix, normal, 90, 90*f*
Arteriography. *See* Angiography.
    CT. *See* Computed tomographic arteriography (CTA).
Arteriosclerotic plaque. *See* Plaque.

Articular cartilage, loss of, in degenerative joint
    disease, 140*f*, 152*f*
Articular disc, of temporomandibular joint, 23*f*
Ascites, 70*f*, 71, 71*f*
Atrial septal defect, 53, 53*f*
Atrium
    left, 49*f*, 50*f*, 54*f*
    right, 39*f*, 50*f*, 54*f*
Auditory canal, external, 28*f*
Avascular necrosis, of femoral head, 141, 141*f*
Axillary lymph nodes, enlarged, 127, 127*f*
Axis (C2), fracture of, 107, 107*f*

B

Baker (popliteal) cyst, 151, 151*f*
Balloon angioplasty, for stenotic coronary
    artery, 51, 51*f*
Basal skull fracture, 28, 28*f*
Benign pleomorphic adenoma, of parotid gland, 25
Benign prostatic hypertrophy, 94, 94*f*
Biceps brachii muscle, 131*f*
Biceps brachii tendon, dislocation of, 128, 128*f*
Biceps sling mechanism, interruption of, 128
Bicornuate uterus, 96, 96*f*
Bifurcation, carotid, plaque at, 16, 16*f*
Bile duct, common
    normal, 87*f*
    tumor obstruction of, 85, 87, 87*f*
Bladder calculus, 100, 100*f*
Bladder diverticulum, 99, 99*f*
Blood, in external auditory canal, 28*f*
Blow-out fracture, 20, 20*f*
Bone. *See also named bone.*
    devascularized fragment of, 141*f*
    lytic tumor of, 138
Bone spurs, 109, 109*f*
Bowel loops
    barium and air in, 74*f*
    in abdominal adenopathy, 72, 72*f*
    in ascitic fluid, 71, 71*f*
Boxer's fracture, 136, 136*f*
Brachiocephalic artery, 57*f*, 61*f*
Brachiocephalic vein, 57, 57*f*
    tumor compression on, 61, 61*f*
Brachioradialis muscle, 131*f*
Brain. *See also specific part.*
    MR images of, 1, 1*f*
Brain stem, 8*f*
Brain tumor
    metastatic, 3, 3*f*
    primary, 4, 4*f*
Breast cancer, mammogram of, 40, 40*f*
Breast cyst, 41, 41*f*

Bursitis
  iliopsoas, 142, 142f
  pes anserinus, 154, 154f

C

Calcaneal tendon tear, 155, 155f
Calcaneus, 155f, 156f, 161f
  fracture of, 156, 156f
Calcified plaque
  at carotid bifurcation, 16, 16f
  in coronary arteries, 49, 49f
  in internal carotid artery, 17, 17f
Calculus
  bladder, 100, 100f
  dilated submandibular duct with, 26, 26f
  renal, 82, 82f
  ureteral, 93, 93f
Cancer. See also specific neoplasm.
  breast
    mammogram of, 40, 40f
    ultrasound of, 41, 41f
  gastroesophageal junction, 75, 75f
  head and neck, 29, 29f. See also specific
     anatomic part.
  lung, 59, 59f
    advanced, 60, 60f
    right upper lobe, 61f, 62
  renal, 83, 83f
Caput medusae, 70, 70f
Cardiomyopathy, hypertrophic, 54, 54f
Carotid artery
  bifurcation of, plaque at, 16, 16f
  common, 17f, 32, 32f, 33f
  external, 17f, 29f
  internal, 17f, 29f
    aneurysm of, 14, 14f, 15, 15f
    soft plaque in, 17, 17f
Carotid siphon, 15f
Cauda equina, 114f
Cecum, 77f, 90f
Cephalhematoma, 2, 2f
Cerebellum, 4f, 7f
Cerebral cortex, atrophied, 11, 11f
Cerebrospinal fluid
  in subarachnoid space, 113f
  in subdural space, 11f
  production of, hydrocephalus associated with, 1
Cervical facet joints
  arthritic, 109f
  degenerative disease of, 109, 109f
  normal, 109f
Cervical intervertebral disc, herniation of, 108, 108f
Cervical lymph nodes
  enlarged, 31, 31f. See also Lymphadenopathy.
  normal-sized, 31f

Cervical vertebrae, bodies of, 109f
Chest wall abnormality(ies)
  pectus carinatum as, 35, 35f
  pectus excavatum as, 36, 36f
Cholelithiasis, 88, 88f
Choroid plexus, 4f
Circumflex artery, 50f
  left, stenosis in, 51, 51f
Cirrhosis, 78, 78f
  caput medusae in, 70, 70f
Clivus, 107f
Coarctation of aorta, 47, 47f
Colitis, ulcerative, 92, 92f
Collateral ligament, medial, tear of, 146, 146f
Colles fracture, 134, 134f
  reversed, 134, 134f
Colon
  rectosigmoid, 92f
  sigmoid, 77f
    volvulus of, 89, 89f
  transverse, 77f
Comminuted fracture
  calcaneal, 156, 156f
  femoral neck, 139f
Common bile duct
  normal, 87f
  tumor obstruction of, 85, 87, 87f
Communicating hydrocephalus, 1
Compression, neural, lumbar disc herniation with,
    118, 118f
Compression fracture, of vertebral body,
    110, 110f
Computed tomographic arteriography (CTA)
  of aberrant right coronary artery, 50, 50f
  of aberrant right subclavian artery, 48, 48f
  of aortic aneurysm
    abdominal, 73, 73f
    ascending, 44, 44f
  of carotid bifurcation plaque, 16, 16f
  of coronary artery disease, 49, 49f
  of internal carotid artery aneurysm, 15, 15f
  of internal carotid artery plaque, 17, 17f
  of obstructed femoral artery, 143, 143f
Computed tomography (CT)
  abdominal
    adenopathy in, 72, 72f
    of ascites, 71, 71f
  of appendicitis, 90, 90f
  of axis fracture, 107, 107f
  of basal skull fracture, 28, 28f
  of benign prostatic hypertrophy, 94, 94f
  of blow-out fracture, 20, 20f
  of calcaneal fracture, 156, 156f
  of caput medusae, 70, 70f
  of cephalhematoma, 2, 2f

Computed tomography (CT) (*Continued*)
  of cervical facet joint degenerative disease, 109, 109*f*
  of cirrhosis, 78, 78*f*
  of coarctation of aorta, 47, 47*f*
  of Crohn's disease, 91, 91*f*
  of deviated nasal septum, 21, 21*f*
  of diaphragmatic hernia, 66, 66*f*
  of dilated submandibular duct, with calculus, 26, 26*f*
  of emphysema, 58, 58*f*
  of enlarged axillary lymph nodes, 127, 127*f*
  of enlarged cervical lymph nodes, 31, 31*f*
  of epidural hematoma, 9, 9*f*
  of esophageal varices, 64, 64*f*
  of foot bones, 160, 160*f*
  of fractured rim of glenoid fossa, 124, 124*f*
  of frontal sinus asymmetry, 19, 19*f*
  of gastroesophageal junction carcinoma, 75, 75*f*
  of hip dislocation with acetabular fracture, 137, 137*f*
  of inguinal hernia, 69, 69*f*
  of internal mammary artery coronary bypass, 55, 55*f*
  of ischemic stroke, 13, 13*f*
  of lung cancer
    advanced, 60, 60*f*
    right upper lobe, 61, 61*f*
  of maxillary and ethmoidal sinusitis, 18, 18*f*
  of mediastinal lymphoma, 43, 43*f*
  of mediastinal tumor, 42, 42*f*
  of meningioma, 12
  of metastatic disease, 67, 67*f*
    acetabular, 138, 138*f*
  of nasal bone fracture, 22, 22*f*
  of pancreatic adenocarcinoma, 85, 85*f*
  of parotid gland tumor, 25, 25*f*
  of pars interarticularis fracture, 111, 111*f*
  of pectus carinatum, 35, 35*f*
  of pectus excavatum, 36, 36*f*
  of pharyngeal mass, 29, 29*f*
  of pleural effusion, 57, 57*f*
  of psoas abscess, 74, 74*f*
  of pulmonary embolism, 39, 39*f*
  of radial head fracture, 130, 130*f*
  of renal calculus, 82, 82*f*
  of renal carcinoma, 83, 83*f*
  of renal cyst, simple, 80, 80*f*
  of sacroiliitis, 117, 117*f*
  of splenomegaly, 79, 79*f*
  of spondylolisthesis, 112, 112*f*
  of subdural hematoma
    acute, 10, 10*f*
    chronic, 11, 11*f*

Computed tomography (CT) (*Continued*)
  of temporomandibular degenerative joint disease, 24, 24*f*
  of teratoma, 98, 98*f*
  of thyroglossal duct cyst, 33, 33*f*
  of ulcerative colitis, 92, 92*f*
  of umbilical hernia, 68, 68*f*
  of ureteral calculi, 93, 93*f*
  of urinary bladder calculus, 100, 100*f*
  of urinary bladder diverticulum, 99, 99*f*
  of vertebral osteomyelitis, 115, 115*f*
  of volvulus, 89, 89*f*
Condyle
  femoral, 145*f*, 147*f*, 149*f*
  mandibular, 24*f*, 28*f*
  tibial, 149*f*
Contusion
  pulmonary, 37
  scalp, 9*f*
Conus medullaris, 120*f*
Coracoid process, 121*f*, 122*f*, 124*f*
Coronary angioplasty, 51, 51*f*
Coronary artery
  anastomosis of internal mammary artery to, 55, 55*f*
  left, 50*f*
  right, 50*f*
    aberrant, 50, 50*f*
Coronary artery disease, 49, 49*f*
Coronary bypass procedures, internal mammary artery, 55, 55*f*
Corpus callosum, 6*f*
Costal cartilage, 55*f*
Costophrenic (costodiaphragmatic) angle, 56*f*
Cranial cavity, anterior fossa of, 19*f*
Cranial nerve (VII), normal, 8*f*
Cranial nerve (VIII)
  normal, 8*f*
  schwannoma of, 8, 8*f*
Cranium, infant, 2, 2*f*
Crohn's disease, 91, 91*f*
Cruciate ligament, anterior
  intact, 150*f*
  tear in, 150, 150*f*
CT. *See* Computed tomography (CT).
CTA. *See* Computed tomographic arteriography (CTA).
Cuboid, 156*f*
Cyst(s)
  Baker (popliteal), 151, 151*f*
  breast, 41, 41*f*
  epididymal, 103, 103*f*
  hepatic, 84*f*
  ovarian, 97, 97*f*
    dermoid, 98, 98*f*

Cyst(s) (*Continued*)
  pineal, 6, 6*f*
  renal
    complex, 81, 81*f*
    simple, 80, 80*f*
  subcortical, degenerative, 140, 140*f*
  thyroglossal duct, 33, 33*f*

D

Deep vein thrombosis, 144, 144*f*
Degenerative cystic changes, 160, 160*f*
Degenerative joint disease
  cervical facet, 109, 109*f*
  hip, 140, 140*f*
  knee, 152, 152*f*
  temporomandibular, 24, 24*f*
Degenerative spondylolisthesis, 113, 113*f*, 114,
  114*f*
Deltoid muscle, 125*f*
Dental metal fillings, imaging artifact due to, 20*f*
Deviated septum, nasal, 21, 21*f*
Diaphragm
  esophagogastric junction at, 63, 63*f*
  in lung cancer patient, 59*f*
    metastatic, 60*f*
  in pleural effusion patient, 56*f*
Diaphragmatic hernia, 65, 65*f*, 66, 66*f*
Diastole, aortic valve during, 52, 52*f*
Disc, intervertebral. *See* Intervertebral disc.
Discitis, infectious, 115, 115*f*
Dislocation. *See also* Fracture(s).
  biceps brachii tendon, 128, 128*f*
  hip, with acetabular fracture, 137, 137*f*
  shoulder, 122, 122*f*, 123, 123*f*
  temporomandibular joint articular disc, 23, 23*f*
Diverticulum
  Meckel's (ileal), 77, 77*f*
  urinary bladder, 99, 99*f*
Duodenal bulb, 76
Duodenal ulcer, 76, 76*f*
Duodenum, 65*f*
Dural sac, 118*f*
  compressed, 119*f*
  fluid in, 110*f*
"Dural tail" sign, of meningioma, 12, 12*f*
Dyspnea, pectus carinatum associated with, 35

E

Effusion
  arthritic facet joint, 113, 113*f*
  knee joint, 145, 145*f*
  pleural, 56, 56*f*, 57, 57*f*
Embolism, pulmonary, 39, 39*f*
Emphysema, 58, 58*f*
  subcutaneous, 37, 37*f*

Enteritis, regional, 91, 91*f*
Epicondyle, medial femoral, 146*f*
Epididymal cyst, 103, 103*f*
Epididymitis, 102, 102*f*
Epidural hematoma, 9, 9*f*
Epiglottis, 29*f*
Epiphyseal growth plates, 153, 153*f*
Erector spinae, 119*f*
Erosions, vertebral body, 115*f*
Esophageal varices, 64, 64*f*
Esophagus, 48*f*
  air in, 47*f*
  tumor constriction of, 75*f*
Esophogram
  air-contrast, of small sliding hiatal hernia,
    63, 63*f*
  radiographic, of aberrant right subclavian
    artery, 48, 48*f*
Ethmoid air cells, orbital contents herniated
  into, 20*f*
Ethmoid sinus, 18*f*, 20*f*
Ethmoid sinusitis, 18, 18*f*
External acoustic meatus, 24*f*
External auditory canal, 28*f*
Extraocular muscles, 20*f*

F

Facet joint
  arthritic
    cervical, 109*f*
    effusion in, 113, 113*f*
  cervical
    degenerative disease of, 109, 109*f*
    normal, 109*f*
  hypertrophic, 114*f*
  lumbar, 118*f*
  normal, 114*f*
Facial nerve (VII), normal, 8*f*
Falciform ligament, 70*f*
Falx cerebri, deviation of
  epidural hematoma causing, 9, 9*f*
  subdural hematoma causing, 10, 10*f*
Fasciitis, plantar, 161, 161*f*
Fat
  periappendiceal, 90, 90*f*
  renal sinus, 83*f*, 93*f*
Fat pads, displaced, 130, 130*f*
Fecalith, calcified appendiceal, 90, 90*f*
Femoral artery
  obstructed, 143, 143*f*
  patent, 143*f*
Femoral condyle, 145*f*, 147*f*, 149*f*
Femoral epicondyle, medial, 146*f*
Femoral head, 99*f*, 137*f*
  avascular necrosis of, 141, 141*f*

Femoral neck
    comminuted fracture of, 139f
    intact, 139f
Femoral shaft, 141f
Femur, 150f
    proximal, fracture of, 139, 139f
Fibroma, uterine (leiomyoma), 95, 95f
Fibula, 153f, 157f
    fracture of, 157f
        distal, 158, 158f
    head of, 154f
Flexor digitorum brevis, 161f
Fluid
    in dural sac, 110f
    in iliopsoas bursa, 142f
    in pleural cavity, 57, 57f
Foot bones. See also specific bone.
    computed tomography of, 160, 160f
Forceps-assisted delivery, cephalhematoma
            associated with, 2
Fossa ovale, intact, 53f
Fracture(s). See also Dislocation.
    acetabular, hip dislocation with, 137, 137f
    ankle, 157, 157f
    axis (C2), 107, 107f
    basal skull, 28, 28f
    blow-out, 20, 20f
    boxer's, 136, 136f
    calcaneal, 156, 156f
    Colles, 134, 134f
    compression, of vertebral body, 110, 110f
    femoral, proximal, 139, 139f
    fibular, 157f
        distal, 158, 158f
    glenoid fossa rim, 124, 124f
    malleolar, medial, 158, 158f
    mandibular, 27, 27f
    nasal bone, 22, 22f
    olecranon, 129, 129f
    pars interarticularis, 111, 111f
    radial head, 130, 130f
    scaphoid, 132, 132f
    Smith, 135, 135f
    tibial, 153, 153f
Frontal sinus
    aerated, 19f
    asymmetry of, 19, 19f
Funnel breast (pectus excavatum), 36, 36f

G

Gallbladder, wall of, 88f
Gallstones, 88, 88f
Gastric antrum, 65f
Gastric cardia, 63, 63f
Gastric fundus, 65f

Gastrocnemius muscle, 151f
Gastroesophageal junction, carcinoma of,
            75, 75f
Genioglossus, 30f
Glenoid, 122f, 125f
    articular surface of, 124f
Glenoid fossa, fractured rim of, 124, 124f
Glenoid labrum, superior, anterior to posterior tear
            of, 126, 126f
Gluteus maximus, 94f, 96f
Goiter, 34, 34f
Gray matter, 9f
Greater trochanter, fracture of, 139
Growth plates, epiphyseal, 153, 153f

H

Hallux
    phalanx of, 160f
    sesamoid bone of, 160, 160f
Hangman's fracture, 107, 107f
Head and neck. See also specific anatomic part.
    cancer of, 29, 29f
Hearing loss, sensorineural, 8
Heart. See also specific part.
    in aortic aneurysm patient, 44f
    in lung cancer patient, 59f
        metastatic, 60f
    in pleural effusion patient, 56f
    in pneumonia patient, 38f
Hematoma
    epidural, acute, 9, 9f
    subdural
        acute, 10, 10f
        chronic, 11, 11f
Hemidiaphragm, dome of, 62f
Hemiscrotal pain, 106
Hepatic. See also Liver.
Hepatic cysts, 84f
Hepatic vein, 70f
Hernia (herniation)
    cervical disc, 108, 108f
    diaphragmatic, 65, 65f, 66, 66f
    hiatal
        large sliding, 62, 62f
        small sliding, 63, 63f
    inguinal, 69, 69f
    lumbar disc, with neural compression, 118, 118f
    umbilical, 68, 68f
Hiatal hernia, sliding
    large, 62, 62f
    small, 63, 63f
Hip
    degenerative joint disease of, 140, 140f
    dislocation of, with acetabular fracture, 137, 137f
    fracture of, 139, 139f

Humeral condyles, 129*f*
Humeral head, 122*f*
Hyaline cartilage, 133, 133*f*
Hydrocele, 104, 104*f*
Hydrocephalus, 1, 1*f*
Hypertrophic cardiomyopathy, 54, 54*f*

I

Ileal (Meckel's) diverticulum, 77, 77*f*
Ileum
    distal, thickened wall of, 91*f*
    terminal, 77*f*
Ileum loops, multiple overlapping, 77*f*
Iliopsoas bursa, fluid in, 142*f*
Iliopsoas bursitis, 142, 142*f*
Iliopsoas muscle, 92*f*, 138*f*, 142*f*
    normal, 74*f*
Ilium, 117*f*
Infarct, ischemic, 13*f*
Infection
    respiratory tract, pectus carinatum
        associated with, 35
    vertebral column, psoas abscess due
        to, 74
Infectious discitis, 115, 115*f*
Inferior nasal concha, 22*f*
Inferior vena cava, 64*f*
Inflammatory bowel disease, 91, 91*f*
Inguinal hernia, 69, 69*f*
Intaorbital air, 20*f*
Interatrial septum, 53*f*
Internal mammary artery, 55*f*
Internal mammary artery coronary bypass,
    55, 55*f*
Intertubercular groove, synovial fluid in, 128*f*
Interventricular septum, 53*f*
    thickened, 54*f*
Intervertebral disc
    cervical
        herniation of, 108, 108*f*
        normal, 109*f*
    collapsed, 115*f*
    lumbar, herniation of, with neural
        compression, 118, 118*f*
Ischemic stroke, 13, 13*f*
Ischioanal fossa, 94*f*
Ischiopubic ramus, 69*f*

J

Joint(s). *See named joint.*
Joint effusion, in arthritic facet joint,
    113, 113*f*
Joint separation, acromioclavicular, 121, 121*f*
Jugular bulb, 8*f*
Jugular vein, internal, 29*f*, 33*f*

K

Kidney, 66*f*. *See also* Renal *entries.*
    transplanted, for polycystic kidney disease,
        84, 84*f*
Knee
    degenerative joint disease of, 152, 152*f*
    joint effusion in, 145, 145*f*

L

Labrum, superior, anterior to posterior tear of, 126,
    126*f*
Lamina papyracea, 20*f*
Leiomyoma (uterine fibroma), 95, 95*f*
Lesser tubercle, 128*f*
Ligament of Treitz, 86*f*
Linear fracture, of petrous part of temporal bone,
    28*f*
Lingual cancer, 30, 30*f*
Liver, 64*f*, 66*f*
    cirrhotic. *See* Cirrhosis.
    metastases to, 67, 67*f*
Liver cysts, 84*f*
Lumbar intervertebral disc, herniation of, with
    neural compression, 118, 118*f*
Lumbar spinal canal stenosis, 119, 119*f*
Lumbar vertebrae, variation in number of, 116,
    116*f*
Lunate, 132*f*, 133*f*
Lung, 55*f*. *See also* Pulmonary *entries.*
    emphysematous, 58*f*
    left, 57*f*
    metastases to, 67, 67*f*
    right, visceral pleura of, 57, 57*f*
Lung cancer, 59, 59*f*
    advanced, 60, 60*f*
    right upper lobe, 61*f*, 62
Lung mass, peripheral, 60*f*
Lymph nodes. *See also specific nodes.*
    in abdominal adenopathy, 72, 72*f*
Lymphadenopathy
    axillary, 127, 127*f*
    cervical, 29, 29*f*. *See also* Cervical lymph nodes.
    mediastinal, 43, 43*f*
Lymphoma, mediastinal, 43, 43*f*
Lytic bone tumor, 138

M

Magnetic resonance imaging (MRI)
    of adult polycystic kidney disease, 84, 84*f*
    of ankle sprain, 159, 159*f*
    of anterior cruciate ligament tear, 150, 150*f*
    of aortic valve, normal, 52, 52*f*
    of aortic valve stenosis, 52, 52*f*
    of atrial septal defect, 53, 53*f*
    of avascular necrosis of femoral head, 141, 141*f*

Magnetic resonance imaging (MRI) (*Continued*)
of Baker (popliteal) cyst, 151, 151*f*
of bicornuate uterus, 96, 96*f*
of brain tumors
metastatic, 3, 3*f*
primary, 4, 4*f*
of calcaneal tendon tear, 155, 155*f*
of cervical intervertebral disc herniation, 108, 108*f*
of degenerative spondylolisthesis, 113, 113*f*, 114, 114*f*
of dislocated biceps brachii tendon, 128, 128*f*
of hydrocephalus, 1, 1*f*
of hypertrophic cardiomyopathy, 54, 54*f*
of iliopsoas bursitis, 142, 142*f*
of ischemic stroke, 13, 13*f*
of knee joint effusion, 145, 145*f*
of lumbar disc herniation, with neural compression, 118, 118*f*
of lumbar spinal canal stenosis, 119, 119*f*
of medial collateral ligament tear, 146, 146*f*
of medial meniscal tear, 147, 147*f*
of meningioma, 12, 12*f*
of patellar tendon tear, 149, 149*f*
of pes anserinus bursitis, 154, 154*f*
of pineal gland cyst, 6, 6*f*
of pituitary microadenoma, 5, 5*f*
of plantar fasciitis, 161, 161*f*
of pronator teres muscle tear, 131, 131*f*
of pseudotumor cerebri, 7, 7*f*
of quadriceps tendon tear, 148, 148*f*
of renal cyst, complex, 81, 81*f*
of rotator cuff tear, 125, 125*f*
of scaphoid fracture, 132, 132*f*
of spinal cord transection, 120, 120*f*
of superior labrum, anterior to posterior (SLAP) tear, 126, 126*f*
of temporomandibular joint dislocation, 23, 23*f*
of tongue cancer, 30, 30*f*
of triangular fibrocartilage complex tear, 133, 133*f*
of uterine fibroma, 95, 95*f*
of vertebral body compression fracture, 110, 110*f*
of vestibulocochlear nerve schwannoma, 8, 8*f*
Malleolus
lateral, 159*f*
medial, 158*f*
fracture of, 158, 158*f*
Malrotation, of small bowel, 86, 86*f*
Mammary artery, internal, 55*f*
Mammography, of breast cancer, 40, 40*f*
Mandible, 25*f*, 29*f*, 107*f*
angle of, 27*f*
fracture of, 27, 27*f*

Mandibular condyle, 24*f*, 28*f*
Mandibular fossa, 24*f*
Mandibular ramus, 30*f*
Manubrium, 35*f*
Mastoid process, 24*f*
Maxillary sinus, 18*f*, 21*f*, 22*f*
orbital contents herniated into, 20*f*
Maxillary sinusitis, 18, 18*f*
Meckel's (ileal) diverticulum, 77, 77*f*
Medial collateral ligament, tear of, 146, 146*f*
Medial meniscal tear, 147, 147*f*
Mediastinal lymphoma, 43, 43*f*
Mediastinal margin, left, 60*f*
Mediastinal mass, metastatic, 60, 60*f*
Mediastinal tumor, 42, 42*f*
Meningioma, 12, 12*f*
Meniscal tear, medial, 147, 147*f*
Mental symphysis, 27*f*
Mesenteric artery, superior, 85*f*
Mesenteric vein, superior, 72*f*, 85*f*
Metacarpal(s)
fifth, 136*f*
fractured, 136*f*
first, 134*f*, 135*f*
Metal, dental, imaging artifact due to, 20*f*
Metastases, 67, 67*f*
acetabular, 138, 138*f*
brain, 3, 3*f*
lung, 60, 60*f*
Metatarsal(s), first, 160*f*
Microadenoma, pituitary, 5, 5*f*
Middle ear, blood in, 28*f*
MRI. *See* Magnetic resonance imaging (MRI).
Mucoperiosteum, thickened, 18*f*
Müllerian duct abnormalities, 96, 96*f*

N

Nasal bone fracture, 22, 22*f*
Nasal concha, inferior, 22*f*
Nasal septum, 21*f*, 22*f*
deviated, 21, 21*f*
Nasomaxillary sutures, vs. nasal bone fracture, 22
Neural compression, lumbar disc herniation with, 118, 118*f*
Neuroectodermal tumor, in parietal lobe, 4, 4*f*
Nodule, thyroid, 32, 32*f*
Non-calcified (soft) plaque
in coronary arteries, 49*f*
in internal carotid artery, 17, 17*f*
Noncommunicating (obstructive) hydrocephalus, 1
Nuclear medicine imaging, 162, 162*f*

O

Olecranon, fracture of, 129, 129*f*
Optic chiasm, 5*f*

Oral pharynx, 30f
Orbit, 21f
    blow-out fracture of, 20, 20f
Orbital floor, 20f
Ossicles, 28f
Ossification, in cephalhematoma, 2
Osteomyelitis, vertebral, 115, 115f
Osteophytes, 109, 109f, 152f
Ostium secundum, 53, 53f
Ovarian cyst, 97, 97f
    dermoid, 98, 98f
Ovary, normal, 97f

P

Pain, hemiscrotal, 106
Pancreatic adenocarcinoma, 85, 85f
Pancreatic parenchyma, normal, 85f
Papilla, sublingual, calculus at, 26f
Papilledema, 7, 7f
Paranasal sinuses. See also named sinus.
    improper drainage of, 21
    pneumatization of, variations in, 19, 19f
Paratracheal lymph node, 43f
Parietal lobe, neuroectodermal tumor in, 4, 4f
Parietal pleura, 57
Parotid gland, tumor of, 25, 25f
Pars interarticularis
    defect of, spondylolisthesis secondary to, 112, 112f
    fracture of, 111, 111f
    intact, 111f
Patella, 148f, 149f
Patella infera, 149
Patellar tendon tear, 149, 149f
Pectus carinatum (pigeon breast), 35, 35f
Pectus excavatum (funnel breast), 36, 36f
Pellegrini-Stieda syndrome, 146
Pes anserinus bursitis, 154, 154f
Pes anserinus tendon, 154f
Phalanx, proximal, of, hallux, 160f
Pharynx
    mass in, 29, 29f
    oral, 30f
Phleboliths, 140, 140f
Pigeon breast (pectus carinatum), 35, 35f
Pineal gland, 6f
    cysts of, 6, 6f
Piriformis, 92f
Pituitary gland, 7f
    tumor of, 5, 5f
Plantar fascia, normal, 161f
Plantar fasciitis, 161, 161f
Plaque
    calcified
        at carotid bifurcation, 16, 16f
        in coronary arteries, 49, 49f

Plaque (Continued)
    non-calcified (soft)
        in coronary arteries, 49f
        in internal carotid artery, 17, 17f
Pleura
    parietal layers of, 57
    visceral layers of, 57, 57f
Pleural cavity, fluid in, 57, 57f
Pleural effusion, 56, 56f, 57, 57f
Pneumonia, 38, 38f
Pneumothorax, 37, 37f
Polycystic kidney disease, adult, 84, 84f
Popliteal (Baker) cyst, 151, 151f
Popliteal vein, thrombus obstructing, 144f
Portal vein, 70f
Postbulbar ulcers, duodenal, 76
Posterior horn, medial meniscus, 147f
Profunda femoral arteries, 143f
Pronator teres muscle, tears of, 131, 131f
Prostate
    benign hypertrophy of, 94, 94f
    normal, 94f
Pseudotumor cerebri, 7, 7f
Psoas abscess, 74, 74f
Psoas muscle, 74f, 82f, 99f
Pubic ramus, superior, 69f
Pubic symphysis, 94f, 100f
    normal, 95f
Pulmonary. See also Lung entries.
Pulmonary artery
    left, in advanced lung cancer patient, 60f
    thrombus in, 39f
Pulmonary contusion, 37
Pulmonary embolism, 39, 39f

Q

Quadriceps tendon, 145f
    tear in, 148, 148f

R

Radiographic studies
    barium-contrast
        of diaphragmatic hernia, 65, 65f
        of small bowel malrotation, 86, 86f
    of acromioclavicular joint separation, 121, 121f
    of ankle fracture, 157, 157f
    of ascending aorta aneurysm, 44, 44f
    of boxer's fracture, 136, 136f
    of Colles fracture, 134, 134f
    of degenerative joint disease of hip, 140, 140f
    of degenerative joint disease of knee, 152, 152f
    of distal fibular fracture, 158, 158f
    of duodenal ulcer, 76, 76f
    of gastroesophageal junction carcinoma,
        75, 75f

Radiographic studies (*Continued*)
 of hip dislocation with acetabular fracture, 137, 137*f*
 of lumbar vertebrae, 116, 116*f*
 of lung cancer, 59, 59*f*
  advanced, 60, 60*f*
 of mandibular fracture, 27, 27*f*
 of Meckel's (ileal) diverticulum, 77, 77*f*
 of medial malleolar fracture, 158, 158*f*
 of olecranon fracture, 129, 129*f*
 of pectus excavatum, 36*f*
 of pleural effusion, 56, 56*f*
 of pneumonia, 38, 38*f*
 of pneumothorax, 37, 37*f*
 of proximal femoral fracture, 139, 139*f*
 of radial head fracture, 130, 130*f*
 of right aortic arch, 46, 46*f*
 of shoulder dislocation
  AP view, 122, 122*f*
  Y view, 123, 123*f*
 of situs inversus, 45, 45*f*
 of sliding hiatal hernia, 62, 62*f*
 of Smith fracture, 135, 135*f*
 of tibial fracture, 153, 153*f*
Radionuclide bone scan, 162, 162*f*
Radioulnar joint, injected contrast material in, 133*f*
Radius, 129*f*, 130*f*, 133*f*
 distal, 132*f*
 head of, fracture of, 130, 130*f*
Rectosigmoid colon, 92*f*
Rectus abdominis, 96*f*
Rectus abdominis muscle, 69, 69*f*
Rectus sheath, 68, 68*f*
Regional enteritis, 91, 91*f*
Renal calculus, 82, 82*f*
Renal carcinoma, 83, 83*f*
Renal collection system, dilated, 93, 93*f*
Renal cyst
 complex, 81, 81*f*
 simple, 80, 80*f*
Renal disease, polycystic, 84, 84*f*
Renal pelvis, calculus in, 82*f*
Renal sinus, fat in, 83*f*, 93*f*
Respiratory tract infections, pectus carinatum
  associated with, 35
Retromandibular veins, 25*f*
Ribs, 57*f*
Rima glottidis, 33*f*
Rotator cuff tear, 125, 125*f*

S

S1 nerve, 118*f*
Sacral body, 117*f*
Sacral foramina, anterior, 117*f*
Sacroiliac joint, 89*f*
 normal, 117*f*

Sacroiliitis, 117, 117*f*
Sacrum, 92*f*
Salivary glands, tumors of, 25, 25*f*
Saphenous vein, great, anastomosis of caput
  medusae with, 70, 70*f*
Scalenus medius, 31*f*
Scalp contusion, 9*f*
Scaphoid, 133*f*
 fracture of, 132, 132*f*
Scapula, lateral border of, 124*f*
Schatzki ring, 63*f*
Schwannoma, vestibulocochlear nerve, 8, 8*f*
Scintigraphy, 162, 162*f*
Scrotum, ultrasound examination of
 epididymal cyst in, 103, 103*f*
 epididymitis in, 102, 102*f*
 hydrocele in, 104, 104*f*
 testicular torsion in, 106, 106*f*
 testicular tumor in, 105, 105*f*
 varicocele in, 101, 101*f*
Sentinel node, 31, 31*f*
Septum, nasal, deviated, 21, 21*f*
Sesamoid bone, of hallax, 160, 160*f*
Shoulder. *See also specific part.*
 dislocation of, 122, 122*f*, 123, 123*f*
Shunt, left-to-right, atrial septal defect resulting, 53
Sialolithiasis, 26, 26*f*
Sigmoid colon, 77*f*
 volvulus of, 89, 89*f*
Sigmoid sinus, 4*f*, 12*f*
"Silhouette" sign, in pneumonia, 38
Sinus(es). *See named sinus.*
Sinusitis, maxillary and ethmoidal, 18, 18*f*
Situs inversus, 45, 45*f*
Skeleton, scintigraphic imaging of, 162, 162*f*
Skull fracture, basal, 28, 28*f*
Sliding hiatal hernia
 large, 62, 62*f*
 small, 63, 63*f*
Small bowel
 malrotation of, 86, 86*f*
 normal, 91*f*
Smith fracture, 135, 135*f*
Sphenoid sinus, 5*f*, 6*f*, 30*f*
Spinal canal, 114*f*
 lumbar, stenosis of, 119, 119*f*
Spinal cord
 cerebrospinal fluid surrounding, 110*f*
 cervical disc herniation impingement on, 108, 108*f*
 complete transection of, 120, 120*f*
Spinous process, 115*f*
Spleen, 64*f*
 abnormal orientation of, 66, 66*f*
 enlarged, 79*f*
 normal, 79*f*

Splenomegaly, 79, 79f
Spondylolisthesis, 112, 112f
    degenerative, 113, 113f, 114, 114f
Sprain, ankle, 159, 159f
Stenosis. *See at anatomic site.*
Stent, in common bile duct, 85, 85f
Sternocleidomastoid muscle, 29f, 33f
Sternum, 35f, 36f
Stomach, 64f. *See also* Gastric *entries.*
    air-fluid level in, sliding hiatal hernia
        and, 62f
Stone. *See* Calculus; Gallstones.
Stroke, ischemic, 13, 13f
Styloid process, 26f
Subarachnoid space, cerebrospinal fluid
        in, 113f
Subclavian artery
    left, 57f
    right, aberrant, 48, 48f
Subcortical cyst, degenerative, 140, 140f
Subcutaneous emphysema, 37, 37f
Subdural hematoma
    acute, 10, 10f
    chronic, 11, 11f
Sublingual papilla, calculus at, 26f
Submandibular duct, dilated, with
        calculus, 26, 26f
Submandibular gland, 30f
Subscapularis muscle, 125f
Subtalar joint, 156f
Superior labrum, anterior to posterior (SLAP) tear,
        126, 126f
Superior pubic ramus, 69f
Superior vena cava, 42f, 43f, 47f
Supraclavicular node, 31, 31f
Supraspinatus tendon, 125f
    tear in, 125, 125f
Symphysis, mental, 27f
Synovial fluid, in intertubercular groove, 128f
Systole, aortic valve during, 52, 52f

T

Talofibular ligament
    normal, 159f
    thickened and torn, 159f
Talus, 156f, 157f, 158f
Tear(s)
    anterior cruciate ligament, 150, 150f
    calcaneal tendon, 155, 155f
    medial collateral ligament, 146, 146f
    medial meniscal, 147, 147f
    patellar tendon, 149, 149f
    pronator teres muscle, 131, 131f
    quadriceps tendon, 148, 148f
    rotator cuff, 125, 125f

Tear(s) (*Continued*)
    superior labrum, anterior to posterior,
        126, 126f
    supraspinatus tendon, 125, 125f
    triangular fibrocartilage complex, 133, 133f
"Teeth," in teratoma, 98, 98f
Temporal bone, petrous part of, 28f
    linear fracture of, 28f
Temporomandibular joint
    articular disc of, 23f
        dislocation of, 23, 23f
    degenerative disease of, 24, 24f
Tentorium cerebelli, 12f
Teratoma, 98, 98f
Testicular torsion, 106, 106f
Testicular tumor, 105, 105f
Testis
    normal, 105f
    normal vessels of, 101f
Thrombosis, deep vein, 144, 144f
Thrombus
    in aortic aneurysm, 73f
    in popliteal vein, 144f
    in pulmonary artery, 39f
Thyroglossal duct cyst, 33, 33f
Thyroid
    enlarged (goiter), 34, 34f
    right and left lobes of, 34f
Thyroid cartilage, 33f
Thyroid isthmus, 34f
Thyroid nodule, 32, 32f
Thyroid tissue, normal, 32f
Tibia, 150f, 153f, 155f, 157f
Tibial condyle, 149f
Tibial epiphysis, distal and proximal, 153f
Tibial fracture, 153, 153f
Tobacco use
    emphysema associated with, 58
    tongue cancer associated with, 20, 30
Tongue, 30f
    carcinoma of, 29, 30, 30f
Torsion, testicular, 106, 106f
Trachea, 32f, 34f, 42f, 43f, 45f, 47f, 48f
    deviation of, 46, 46f
    in lung cancer patient, 61f
Transplantation, kidney, for polycystic kidney
        disease, 84, 84f
Triangular fibrocartilage complex (TFCC) tear, 133,
        133f
Triquetrum, 132f, 133f
Tumor(s). *See also* Cancer; *specific tumor.*
    bone, lytic, 138
    brain
        metastatic, 3, 3f
        primary, 4, 4f

Tumor(s) (*Continued*)
    mediastinal, 42, 42f
    parotid, 25, 25f
    pituitary, 5, 5f
    testicular, 105, 105f

U

Ulcer, duodenal, 76, 76f
Ulcerative colitis, 92, 92f
Ulna, 129f, 130f, 133f
    head of, 135f
Ultrasonography
    of breast cancer, 41, 41f
    of breast cyst, 41, 41f
    of common bile duct obstruction, 87, 87f
    of deep vein thrombosis, 144, 144f
    of epididymal cyst, 103, 103f
    of epididymitis, 102, 102f
    of gallstones, 88, 88f
    of goiter, 34, 34f
    of hydrocele, 104, 104f
    of ovarian cyst, 97, 97f
    of renal carcinoma, 83, 83f
    of testicular torsion, 106, 106f
    of testicular tumor, 105, 105f
    of thyroid nodule, 32, 32f
    of varicocele, 101, 101f
Umbilical hernia, 68, 68f
Umbilicus, 68, 68f
Ureter
    calculi in, 93, 93f
    normal, 93f
Urinary bladder calculus, 100, 100f
Urinary bladder diverticulum, 99, 99f
Urolithiasis, 82, 82f, 93, 93f, 100, 100f
Uterine fibroma (leiomyoma), 95, 95f
Uterus
    bicornuate, 96, 96f
    normal, 95f

V

Vagina, normal, 95f
Valvular stenosis, aortic, 52, 52f
Varices, esophageal, 64, 64f
Varicocele, 101, 101f
Vena cava
    inferior, 64f
    superior, 42f, 43f, 47f
Venous thrombosis, deep, 144, 144f
Ventricle (cardiac)
    left, 45f, 46f, 55f
        posterolateral wall of, 54f
    right, 54f
Ventricle (cerebral), 1f, 2f, 3f
Vertebrae
    lumbar, variation in number of, 116, 116f
    osteomyelitis of, 115, 115f
Vertebral artery, 31f, 107f
Vertebral body
    cervical, 109f
    compression fracture of, 110, 110f
    erosions in, 115f
Vertebral column infection, psoas abscess due
        to, 74
Vestibulocochlear nerve (VIII)
    normal, 8f
    schwannoma of, 8, 8f
Visceral pleura, of right lung, 57, 57f
Volvulus, 89, 89f

W

White matter, 9f

X

Xiphoid process, 35f, 36f

Z

Zollinger-Ellison syndrome, 76
Zygapophyseal joints. *See* Cervical facet joints.

Printed and bound by CPI Group (UK) Ltd, Croydon, CR0 4YY

03/10/2024

01040451-0005